JOHN WESLEY POWELL

John Wesley Powell at his desk, ca. 1894 (*National Anthropological Archives, Smithsonian Institution*)

JOHN WESLEY POWELL

An Annotated Bibliography

Marcia L. Thomas

Bibliographies and Indexes in American History, Number 49

PRAEGER

Westport, Connecticut
London

Library of Congress Cataloging-in-Publication Data

Thomas, Marcia L., 1950–
John Wesley Powell : an annotated bibliography / Marcia L. Thomas.
p. cm. — (Bibliographies and indexes in American history, ISSN 0742–6828 ; no. 49)
Includes bibliographical references and index.
ISBN 0–313–31942–1 (alk. paper)
1. Powell, John Wesley, 1834–1902—Bibliography. 2. Colorado River (Colo.–Mexico)—Discovery and exploration—Bibliography. 3. Grand Canyon (Ariz.)—Discovery and exploration—Bibliography. 4. West (U.S.)—Discovery and exploration—Bibliography. I. Title. II. Series.
Z8709.5.T47 2004
[F788.P68]
917.91′3—dc22 2004042286

British Library Cataloguing in Publication Data is available.

Library of Congress Catalog Card Number: 2004042286
ISBN: 0–313–31942–1
ISSN: 0742–6828

First published in 2004

Praeger Publishers, 88 Post Road West, Westport, CT 06881
An imprint of Greenwood Publishing Group, Inc.
www.praeger.com

Printed in the United States of America

The paper used in this book complies with the Permanent Paper Standard issued by the National Information Standards Organization (Z39.48–1984).

10 9 8 7 6 5 4 3 2 1

To Madeleine and Christopher.

Contents

Acknowledgments

Bibliographers are eternally indebted to libraries and their staff. Special thanks are due to my colleagues at The Ames Library at Illinois Wesleyan University for their support and assistance. I am especially grateful for Tony Heaton's ability to locate the most obscure sources. Archivist Anke Voss-Hubbard provided expert advice on primary sources and the organization of this bibliography. Administrators and colleagues at Illinois Wesleyan University were most generous in granting me a leave of absence and financial assistance. The quality of service from the librarians and staff at the National Anthropological Archives, the Huntington Library, the University of Arizona, the Manuscript Division of the Library of Congress, the National Archives, and the Utah State Historical Society was uniformly excellent. I am grateful for special assistance from Richard Green at the Smithsonian Museum of Natural History Library, James Haug at the Smithsonian Institution's Department of Anthropology Library, John Irwin at the Denver Public Library, and Gregory Finnegan at Harvard's Tozzer Library.

For this publication Martin White created a splendid set of indices. Copy editor Stacey Shimizu provided knowledge, expertise, and a great deal of patience as we worked through organizational complexities and the peculiarities of 19th-century publishing. To Greenwood Press editor Brien McDonald, I send chocolate. For moral support and for sharing my passion for this work, I am grateful to Earle Spamer at the Academy of Natural Sciences, Richard Quartaroli at the University of Northern Arizona's Cline Library, and Bill Steinbacher-Kemp at the McLean County (Illinois) Museum of History.

Finally, words are inadequate to fully express my appreciation and admiration for two individuals who inspired and supported me throughout this project. To Donald Worster, thank you for so generously opening many doors and believing in the value of this work. I cannot image a kinder, wiser scholar and human being. And to Anders Dahlgren, my traveling companion, technical advisor, proofreader, and dearest friend, thank you for your unwavering confidence.

Introduction

JOHN WESLEY POWELL: AN OVERVIEW

Nineteenth-century American explorer and national hero John Wesley Powell has been a subject of interest to scholars and the general public for nearly 150 years, but recently there has been a flurry of books and articles devoted to the one-armed adventurer. Once best known for leading the first successful expedition through the Grand Canyon in 1869, Major Powell is now recognized for his contributions to the ethnographic study of the indigenous people of the American Southwest and for his views on land use in the arid West. Indeed, his grasp of complex environmental issues and his controversial proposals for reforming public land laws is at the heart of this recent revival.

A study of Powell's work yields an astonishing array of contributions to geology, geography, anthropology, ethnology, environmental history, Native American studies, philosophy, and natural science. His biography is equally impressive. Before he ever set one foot west of the Mississippi to explore the last uncharted territory in the American West, Powell enlisted as an Illinois volunteer at the start of the Civil War. Two years later, he lost his right forearm to a Confederate bullet in the Battle of Shiloh. Yet this slight, scrappy man scrambled up and down the steep walls of the Colorado River canyons like a mountain goat, pushing himself and his men to the limits of their physical and emotional endurance.

Powell spent the next decade surveying and mapping the last blank place on the map of the United States before moving to the nation's capital to embark on a distinguished career as a federal bureaucrat. But the land and people of the American Southwest never relinquished their hold on his imagination. Inspired by the strangely beautiful landscape of the canyon country of Arizona and Utah, this keen observer conceptualized and named several geological processes, such as *base line of erosion* and *antecedent valley*. Spurred by his encounters with the native peoples of the Colorado River Plateau and his proclivity to classify, Powell played a key role in establishing the discipline of anthropology in North America. He collected hundreds of Indian artifacts, as well as words, stories, and myths, for the Smithsonian Institution. He developed a system of classification for language stocks that stands to this day with only minor revision. For his time, Powell's views on American Indians were sympathetic and progressive. He believed that in the interest of their survival, they should be given education, land, and tools, to help them adapt to their inevitable and final domination.

Throughout his years in Washington, Powell was a member and leader of numerous professional organizations, such as the National Academy of Sciences and the American Association for the Advancement of Science, and was instrumental in founding such intellectual organizations as the Anthropological

Society of Washington, the Cosmos Club, and the National Geographic Society, to name but three. His organizational skills and deeply held faith in science ultimately played out in his simultaneous tenure as Director of the Smithsonian Institution Bureau of Ethnology (1879–1902) and Director of the U.S. Geological Survey (1881–1894). He worked tirelessly, lobbying Congress and promulgating the practical benefits of government science. And he was successful—until he attempted to shape his observations on the arid lands of the West into public policy that directly clashed with the special interests of homesteaders, land investors, and powerful corporations. As early as 1878, Powell had stated in *Report on the Lands of the Arid Region* that "the right to the water should adhere in the land where it is used ... not to the individual or company constructing the canals." By 1890, he added an even more radical notion: use of the land was so limited by available water that the public lands should be held from settlement until they could be surveyed. In his attempt as director of the U.S. Geological Survey to complete an irrigation survey, Powell fell hard and fast from political favor. When he resigned his post as head of the survey in 1894, his career and health quickly faded. By 1902, he was dead.

ABOUT THIS BIBLIOGRAPHY

Scope

The writings of John Wesley Powell reflect not only his own personal and professional interests, but also issues of national importance during the transformative years from the end of the Civil War through the domestication of the Western frontier. As Wallace Stegner notes in his highly acclaimed biography of Powell, *Beyond the Hundredth Meridian*, "Because Powell's work proliferated in so many directions and influenced in so many ways so many kinds of people, not only during his lifetime but since, a proper list of writings consulted in the preparation of his biography would be enormous. It would ideally cover everything of importance on the physical history and development of the West after the Civil War; and through its extension into the history of exploration, of the Indian, of the earth sciences, and of irrigation and reclamation, it could be indefinitely lengthened back into the past and on into the future" (369).[1]

While not "indefinitely" exhaustive as the scope imagined by Stegner, this annotated bibliography encompasses Powell's articles in popular magazines and scientific journals, published and unpublished administrative records and correspondence, government publications, Congressional testimony and public addresses, monographs ranging in size from pamphlets to a lengthy philosophical treatise, collections of archival and manuscript materials, and even

[1] Stegner, Wallace. *Beyond the Hundredth Meridian: John Wesley Powell and the Second Opening of the West*. Boston: Houghton Mifflin, 1954.

a few poems. Microform editions are not referenced, although availability of microfilm is noted in annotations for archival material.

Several members of Powell's expeditions and survey teams published their journals and letters, which are included in the section on published and unpublished archival and manuscript collections. These are important not only for an understanding of Powell's personality, but for the light they shed on the controversies surrounding the 1869 and 1871–1872 Colorado River expeditions. When Powell merged incidents from both into the single narrative known as *Exploration of the Colorado River of the West*, omitting any mention of the individuals in the second party, he incurred the wrath of several of its members (see notes to entry #242). In addition, scholars and amateur historians alike continue to debate the affair of Separation Rapids in August 1869. Here, just two days before reaching the end of their perilous journey, brothers Oramel and Seneca Howland and William Dunn left the Powell expedition in 1869. Whether they left because they were cowards or because Powell forced them to, the three men were never seen alive again.

This bibliography is also an annotated compilation of selected references to published biographical material, articles, essays, nonprint material, Internet resources, and monographs about Powell's life and work. Almost as fascinating and perplexing as John Wesley Powell himself is the wide range of reaction to him during his life and in the century following his death. Public and scholarly interest in Powell has fluctuated through the decades, peaking during the centennial of the 1869 Grand Canyon voyage and again on the centennial of his death. One need only compare the works of his principal biographers and critics—Martha D. Lincoln, Grove Karl Gilbert, Robert Stanton, William Culp Darrah, Wallace Stegner, and Donald Worster—to understand Mike Foster's assessment of Powell in *Strange Genius*: "This brilliant thinker, this energetic activist, this subtle bureaucrat, this slick salesman, this idealistic and devious man, this richly complicated man" (202).[2] Beyond the controversies surrounding his actions on the Colorado River, there are important social and political issues connected to Powell: the fate of the indigenous people of North America, the role of the federal government in the affairs of western states, conservation of natural resources, and historical interpretation of the settlement of the American West.

Organization and Coverage

In addition to connecting the reader, student, or scholar to essential works by and about John Wesley Powell, my intent is to bring order to the literature, given the vagaries of 19th-century publishing, the complexities of government publications, and Powell's penchant for taking every opportunity to publish in multiple venues. With few exceptions (such as self-explanatory titles), references are fully annotated. Annotations are informative rather than critical,

---

[2] Foster, Mike. *Strange Genius: The Life of Ferdinand Vandiveer Hayden*. Niwot, CO: Roberts Rinehart, 1994

to indicate the nature of the content or significance of the author. Annotations for works by Powell are longer and more detailed than those for most secondary sources and often paraphrase rather than summarize. All annotations are based on a reading of the work itself or an edition of the work, except for where I indicate that I have not seen an item.

This bibliography is organized in two basic arrangements. Works by Powell (Chapter 2) are chronological and subdivided by type of publication. The remaining chapters are organized alphabetically by author or title, and some are subdivided by topic or chronology. Numerous citations might logically appear in more than one category, thus the Subject Index is especially important. The Author Index and the Title Index for works by Powell, along with cross-references within annotations, also aid the researcher in making associations among specific works or works published in different editions or with variant titles. In all cases, cross-reference numbers in annotations and numbers following index entries point to citation numbers.

Chapter 1 serves as a guide to significant archival and manuscript collections or published works considered to be primary sources for historical study, such as memoirs and published diaries. Powell's closest friend, Grove Karl Gilbert, destroyed Powell's personal papers immediately following his death, at Emma Powell's request. Gilbert kept some items, but his collection in turn was dispersed to private hands. Because personal papers are lacking, I have included accessions with unique material or several letters by Powell. However, I have not attempted to identify the location of every missive, in part because copies of hundreds of outgoing letters from Powell's administration in the Bureau of American Ethnology and the U.S. Geological Survey are available in federal repositories. Many of his other writings are scattered throughout the country in libraries, archives, universities, and private collections. While not comprehensive, the sections containing published and unpublished manuscripts of Powell's associates are particularly thorough. Collections of associated archival material (collections of photographs, maps, and digital material) are included in Chapter 7.

Chapter 2, comprising works by Powell, forms the heart of this bibliography. My intent is to cite all of Powell's published books, pamphlets, and journal articles, including full texts of speeches. I chose selectively in citing extracts, remarks, and abstracts of speeches. Included are chapters, prefaces, forewords, and introductions by Powell contained in monographs written by others. From the enormous body of Powell's testimony, statements, and remarks published in Congressional and administrative records, I cite the most significant works, such as his annual director's reports for the Bureau of American Ethnology and the Geological Survey. In this category are whole works, including letters and excerpts from testimony published in the United States Serial Set. A few of the most significant articles from major newspapers are included if they contain extensive quotations of remarks by Powell or summarize an important public address.

One of the major challenges in compiling this bibliography was devising a logical system for presenting the many reprints and offprints that were published in the same year and in multiple venues (such as an article published in conference proceedings, journal article, and pamphlet edition of the text of an address). It can be exceedingly difficult, if not impossible, to determine the first publication of a work. See, for example, entries for "Mythologic Philosophy" (#272, #277, #280) also published as "Sketch of the Mythology of the North American Indians" (#290, #298). I therefore chose not to cite a work as a reprint unless it was published at a much later date.

The final section in Chapter 2 contains reprints of whole works by Powell that were published after his death. The many textbooks and anthologies that contain short excerpts from *Exploration of the Colorado River of the West* or *Canyons of the Colorado* fall outside the scope of this bibliography. The reader should also pay special attention to cross-references within the annotations, because Powell frequently "borrowed" texts from himself. A good example is "Statement of the Organization, Business Methods and Work of the United States Geological Survey" (#379).

The chapters containing works about Powell and his associates are selective but thorough. Selections are especially representative of published magazine and journal articles that appeared in Powell's time. References to 19th-century works not only are more difficult to locate, but also demonstrate Powell's influence on his contemporaries and their reactions to him. References to 20th- and 21st-century works are more selective and include articles, biographies and obituaries, a few encyclopedia entries and review essays, and monographs with substantial sections on Powell. Contemporary textbooks, general histories with only brief references to Powell, and similar publications are excluded. A representative selection of books and articles intended for a popular audience demonstrates his broad public appeal as a national heroic figure as well as the ire he still evokes for his views on water rights and his management of the Grand Canyon expeditions. I include some background information, such as histories of the U.S. Geological Survey and biographies of political opponents. An understanding of the social, political, and intellectual context of the last half of the 19th century is especially important for a thorough understanding of Powell's life and work.

Chapter 5 contains a representative sample of divergent views of historians and critics published during and following the centennial of the 1869 Colorado River expedition. The most intense interest is focused on his views regarding public land policy, irrigation, and land use in the arid West. Several authors note that Powell's life and work have been reinvented and reinterpreted in numerous ways through the decades.

The coverage of material in the final two chapters is judicious and includes theses and dissertations, works for a juvenile audience, and fictional accounts of events in Powell's life. Internet resources cannot be ignored. Amidst amateur, inaccurate Web pages, one can find a wealth of invaluable digital reproductions of photographs, texts, and manuscripts from universities and libraries. The sections on cartographic and photographic collections are highly selective, but provide important supplemental material because of Powell's role in the surveying and mapping of the American West, as well as his skillful use of photographers and artists to promote the Powell Survey and document vanishing Native American cultures.

PROBLEMS WITH THE POWELL CANON

Powell's fascination with language led to some idiosyncrasies in his writing that present special challenges to the bibliographer. He loved to invent words, such as *intellection*. He disliked the terminal *-al* in adjectives, preferring *geologic* to *geological*. I have retained Powell's spelling in titles and in quotations from his text. The same is true for technical terms, names of Indian tribes, and other words inconsistently spelled, such as *encyclopaedia/ encyclopedia* and *Pai-Ute/Paiute*. Publisher and journal titles, too, have changed over time. I have used such titles and names as they appear in the work at hand. Some titles are abbreviated (see list of abbreviations following this introduction).

Titles of government publications present a different challenge. Many official reports are preceded by letters of transmittal from an agency head. For example, *Geological and Geographical Surveys. Letter From the Secretary of the Interior, Transmitting a Report of Professor Powell in Regard to Surveys* (#259) may be cataloged in libraries under the title "Letter from the Secretary of the Interior...." Whenever possible, I have chosen to use the unique title of the report as printed in the document, relegating transmitting letters to subtitle status.

Three of Powell's works are mentioned so frequently that their cross-referenced numbers are not repeated in the annotations. They are *Exploration of the Colorado River of the West and Its Tributaries* (abbreviated as *Exploration* in annotations), *Report on the Lands of the Arid Regions of the United States* (abbreviated as *Report on the Lands* in annotations), and *Canyons of the Colorado*.

Anyone studying Powell is advised to understand the histories of the Bureau of American Ethnology and the Geological Survey. The survey directed by Powell was known as "The Geographical and Geological Survey of the Rocky Mountain Region Survey" from 1876–1879, but from 1874 to 1876 its official title was "The Geological and Geographical Survey of the Territories, Second Division." Powell's earlier surveys have no official title and were referred to by Congress as "The Survey of the Colorado of the West and Its Tributaries" and administered by the Smithsonian Institution. In 1874 Powell was authorized to continue his work under the direction of the Secretary of the Interior. Whatever its official name, the Powell Survey was officially dissolved when the Bureau of Ethnology and the Geological Survey were created in 1879, but Powell's work continued under his leadership of both agencies. The Bureau of Ethnology changed its name to Bureau of American Ethnology in 1897. In 1964, the bureau merged with the Division of Ethnology of the National Museum to form the Office of Anthropology of the Smithsonian.

ESSENTIAL WORKS

Without question, Powell's most important works are *Exploration of the Colorado River of the West and Its Tributaries* and *Report on the Lands of the Arid Region of the United States*. These two government reports have been studied, reprinted, and quoted since they were published. On the subject of arid lands, his second annual report of the Irrigation Survey in Part II of the *Eleventh Annual Report of the United States Geological Survey, 1890–91* (#419), is significant because it brings together many of his statements and written reports on the subject of settlement and water use in the West. The three articles on irrigation published in *Century Magazine,* "The Irrigable Lands of the Arid Region," "The Non-irrigable Lands of the Arid Region," and "Institutions for the Arid Lands" (#400, #401, #406), should also be read for a full understanding of Powell's views. In the areas of ethnology and anthropology, the reader might start with the *First Annual Report of the Bureau of Ethnology, 1879–80* (#293). In his first director's report, Powell lays out his plan for the work of the bureau. The four accompanying papers in the annual report express his thoughts on topics dear to his heart: "On the Evolution of Language" (#297), "Sketch of the Mythology of the North American Indians" (#298), "Wyandot Government: A Short Study of Tribal Society" (#300), and "On Limitations to the Use of Some Anthropologic Data" (#295). For a thoughtful compilation of Powell's writings and intelligent commentary, consult *Seeing Things Whole: The Essential John Wesley Powell,* by William deBuys (#213). See James Aton's monograph, *John Wesley Powell* (#912), for an analysis of *Exploration* as a work of literature and a critique of Powell's work and writings.

The first biography of Powell was published immediately following his death as a six-part series in the journal *Open Court* (#621 #622, #645–#648). Written by Martha D. Lincoln and Grove Karl Gilbert, these articles, along with William Davis' eulogy presented to the National Academy of Sciences (#611), formed the basis for biographical publications through the first half of the 20th century. In the first book-length biography, *Powell of the Colorado* (#609) published in 1951, William Culp Darrah made the first attempt to document Powell's life. Wallace Stegner's biography, *Beyond the Hundredth Meridian,* received critical acclaim for the power of the author's writing and passion for his subject. Donald Worster's scholarly biography, *A River Running West* (#689), was published in 2003 and stands today as the definitive work on Powell.

No student of Powell can do without volumes 15 and 16–17 of the *Utah Historical Quarterly* (#102, #103). Together, they bring together biographical sketches and original documents from the Colorado River expeditions of 1869 and 1871–72. Another important source, available at some libraries, is a 10-reel set of microfilmed correspondence from the U.S. Geological Survey Record

Group 57 (#7): Powell Survey Letters Received 1869–1879, set number M156. An excellent source of information for Powell's anthropological work is Don and Catherine Fowler's *Anthropology of the Numa: John Wesley Powell's Manuscripts on the Numic Peoples of Western America, 1868–1880* (#29).

Researchers interested in Powell's recorded testimony and statements before Congress and its committees should consult the index to the Congressional Record. Other sources not generally included in this bibliography are newspaper articles, with the exceptions noted previously. *The New York Times* and the *Chicago Tribune* are widely available in libraries. The Denver Public Library and the Utah State Historical Society have good indices of 19th-century newspapers published in their respective states.

SOURCES CONSULTED

This work is based on the last nearly comprehensive bibliography of Powell's writings, compiled by P. C. Warman, Chief of the Editorial Division of the U.S. Geological Survey, and published in 1903. "Catalogue of the Published Writings of John Wesley Powell" (#980), while useful to biographers and scholars of the past century, is incomplete and contains duplicates. Many of the entries, which are topically organized, are briefly annotated or list subtitles; descriptive information is often included. In this work, I have corrected Warman's errors, eliminated duplication, and added new references. In his annotations, Warman indicates reprints and offprints of articles in notes, such as "Issued in separate form, with paper cover bearing title." I cite these "separates" as unique entries. For those I could not verify or locate in library catalogs or finding aids, my annotation states, "Warman separate not seen."

Numerous scholarly monographs and dissertations include useful but incomplete lists of published works by or about John Wesley Powell, and I have consulted many of the titles contained herein. Donald Worster's *A River Running West* (#689) is very useful for secondary sources. *Anthropology of the Numa*, by Don D. and Catherine S. Fowler (#29), has an excellent bibliography. Schmeckebier's *Catalogue and Index of the Publications of the Hayden, King, Powell, and Wheeler Surveys* (#976) is a good source for publications of the early Western surveys. For other published bibliographic sources, see the section "Bibliographies and Guides" in Chapter 5.

Special mention needs to be given to Earle Spamer's Web-based *Bibliography of the Grand Canyon and Lower Colorado River* (#977). This comprehensive, searchable bibliography provided me with numerous citations to obscure works and reprints. Anyone with an interest in Powell and the Southwest must consult this impressive work-in-progress.

Hundreds of citations for secondary sources about Powell, his time, and such related subjects as environmental policy can of course be found in numerous electronic databases, such as WorldCat and RLIN, and in online catalogs, such as those for the Smithsonian Institution (SIRIS), New York Public Library (CATNYP), and the Utah State Historical Society.

FINAL COMMENTS

A truly exhaustive bibliography of writings by and about John Wesley Powell is probably not humanly possible—at least by this human. I have attempted to assemble here a comprehensive list of works *by* Powell; to offer a select list of the best of the writings *about* Powell; and to organize and make accessible these essential materials for the scholar, the enthusiast, and the amateur.

I know, though, that there will be materials overlooked, new publications, and errors to be corrected. My intent is to assemble emendations and supplementary references into a Web-based addendum to this bibliography. With this in mind, I welcome corrections, additions, and comments from readers of this work (powellbibliography@yahoo.com).

List of Abbreviations

General

AAAS	American Association for the Advancement of Science
APS	American Philosophical Society
ARC	*Archival Research Catalog*
ASW	Anthropological Society of Washington
BAE	Bureau of American Ethnology
BSW	Biological Society of Washington
GPO	Government Printing Office
ISU	Illinois State University
IWU	Illinois Wesleyan University
LC	Library of Congress
NAA	National Anthropological Archives
NARA	National Archives and Records Administration
NAS	National Academy of Sciences
NYPL	New York Public Library
PSW	Philosophical Society of Washington
RG	Record Group
Rpt.	Reprint
RU	Record Unit
SIA	Smithsonian Institution Archives
SMC	*Smithsonian Manuscript Collection*
UHQ	*Utah Historical Quarterly*
USGS	U.S. Geological Survey
USNM	U.S. National Museum
Washington	Washington, D.C.

Congressional Series

Cong.	Congress
Doc.	Document
Ex.	Executive
H.	House
Misc.	Miscellaneous
Rept.	Report
S.	Senate

Chronology

1830	Parents Joseph and Mary Dean Powell emigrate from England to New York
1834	John Wesley Powell born at Mount Morris, New York, on March 24
1838	Powell family moves to Jackson, Ohio; studies natural history with George Crookham
1846	Powell family moves to South Grove in southwest Wisconsin
1851	Powell family moves to Bonus Prairie in northwest Illinois; John Wesley helps manage the family farm
1852	Teaches in country school
1853–1858	Intermittently attends college (Illinois Institute, Illinois College, and Oberlin College); teaches in country schools; travels the Mississippi and Ohio river valleys and the Great Lakes, collecting plants and fossilized mollusks
1854	Family moves to Wheaton, Illinois
1858	Joins the Illinois Natural History Society as curator of conchology
1859	Teacher and superintendent of schools in Hennepin, Illinois; continues summer collecting expeditions
1860	Lectures on geography and geology in Tennessee, Kentucky, and Mississippi
1861	Enlists as private in the 20th Illinois Infantry Regiment, Company H; mustered into service as captain of Battery F, 2nd Light Artillery, Illinois Volunteers; marries half-cousin, Emma Dean, on November 28
1862	Right arm is amputated below the elbow following injury at the Battle of Shiloh on April 6
1863	Returns to active duty, with his wife following; commands Battery F at the siege of Vicksburg; awarded honorary MA from Illinois Wesleyan University
1864	Trains freed slaves in weapon use; promoted to major; erects defenses at the Battle of Nashville
1865	Honorably discharged from service with rank of brevet lieutenant colonel in January

1865–1868	Holds position of professor of natural sciences at Illinois Wesleyan University in Bloomington
1866–1872	Holds position of professor at Illinois State Normal University (taught one year only)
1867	Appointed curator of the Illinois Natural History Museum in Normal in March; leads a summer research and collecting expedition of students and amateur scientists (including Emma Dean Powell) to the Colorado Rocky Mountains, resulting in his first published report, "Scientific Expedition to the Rocky Mountains"
1868	Leads second research expedition to Colorado Rockies, with excursions to the Grand, Green, and Yampa Rivers; succeeds in first recorded ascent of Long's Peak
1869	Leads a party of 10 men in four boats down the Green and Colorado Rivers and through the Grand Canyon (the first recorded successful navigation), May 24–August 30
1870	Lectures on the 1869 expedition in western and midwestern cities in spring; receives Congressional appropriation to direct exploration of the Colorado River Plateau north of the Grand Canyon, undertaken in summer and fall (Powell Survey).
1871	Begins second expedition down the Green and Colorado Rivers, May 22; daughter Mary Dean born in Salt Lake City, September 8
1872	Second expedition continues, with ethnological fieldwork and partial descent through Grand Canyon ending September 9; Powell resigns museum curatorship; sells home in Normal and moves family to Washington, D.C.
1873	Appointed Special Commissioner with G. W. Ingalls to investigate condition of Indians in Utah and eastern Nevada; visits bands of Indians, conducts fieldwork in Utah Territory
1874	Publication of *Report of Special Commissioners J. W. Powell and G. W. Ingalls on the Condition of the Ute Indians of Utah;* geological fieldwork continues in eastern Utah
1875	Publication of *Exploration of the Colorado River of the West and Its Tributaries*
1876	Publication of *Report on the Geology of the Eastern Portion of the Uintah Mountains*. Powell Survey officially authorized as U.S. Geological and Geographical Survey of the Territories in the Department of the Interior; fieldwork continues

1877 Publication of *Introduction to the Study of Indian Languages*, 1st ed. Awarded honorary PhD from Illinois Wesleyan University; delivers lecture "The Public Domain" at the spring meeting of the National Academy of Sciences

1878 Publication of *Report on the Lands of the Arid Region of the United States*. Formally recommends to the National Academy of Sciences that federally funded surveys be consolidated into one agency, published as *Report on the Methods of Surveying the Public Domain*; last year of fieldwork in Utah; organizes the Cosmos Club

1879 Powell is instrumental in the establishment of both the Smithsonian Institution's Bureau of Ethnology (later renamed Bureau of American Ethnology) and the U.S. Geological Survey; legislation consolidates the King, Hayden, Wheeler, and Powell Surveys into the USGS; Powell appointed first director of the Bureau of Ethnology (1879–1902); founds the Anthropological Society of Washington; appointed by President Hayes to the Public Lands Commission, charged with recommending revisions to public land laws

1880 Publication of *Introduction to the Study of Indian Languages*, 2nd ed. Clarence King appointed first director of USGS

1881 Appointed by James Garfield as second director of the USGS (1881–1894)

1884 Testifies before the Allison Commission on behalf of government science and activities of the USGS; undertakes plan for topographical mapping of the United States by the USGS

1885 Continues testimony before the Allison Commission; congressional opponents attempt but fail to cut funding for USGS

1886 Awarded honorary degrees by Harvard University and the University of Heidelberg

1888 Congress authorizes the USGS to investigate water sources for development of irrigation in the West (Irrigation Survey); National Geographic Society founded, with Powell as charter member

1889 Accompanies Senate Select Committee on Irrigation on an inspection tour of the West; lectures in North Dakota, Montana, and Nebraska on local control of water resources; serves as president of the American Association for the Advancement of Science; Department of Interior withdraws all public domain from entry

1890	Publication of "The Irrigable Lands," "The Non-Irrigable Lands," and "Institutions for the Arid Lands" in *Century Magazine*. Congress cuts funds for Irrigation Survey
1891	Publication of "Indian Linguistic Families of America North of Mexico" and accompanying map, *Linguistic Stocks of American Indians*, in the 7th annual report of the Bureau of American Ethnology. Powell receives Cuvier Prize from Academie des Sciences, Paris
1892	Senator William Stewart leads successful attack on appropriations for USGS
1893	Powell delivers controversial address on insufficiency of water in the West to the International Irrigation Congress in Los Angeles, October 13
1894	Powell resigns as director of USGS, citing poor health; surgery performed on amputated arm
1895	Publication of *The Canyons of the Colorado* and the first three monographs of the National Geographic Society, *Physiographic Processes, Physiographic Features, Physiographic Regions of the U.S.*
1896	Publication of *The Geologic Map of the United States* begins; Powell moves summer residence to Haven, Maine
1898	Publication of *Truth and Error, or the Science of Intellection*
1900	Tours Cuba and Jamaica with William Henry Holmes, Arnold Hague, and Samuel Langley; health declines rapidly
1902	Dies at home in Maine, September 23

Chapter 1

Manuscript and Archival Sources

A. MANUSCRIPT AND ARCHIVAL SOURCES FOR JOHN WESLEY POWELL

For selected collections of photographs, maps and digital archives, see Chapter 7.

1. Major Collections

1. Henry E. Huntington Library, Papers of Otis R. Marston, 1870–1978. San Marino, CA.

"Dock" Marston was a river runner and historian of the Colorado and Green River basins. This collection of personal and professional papers includes a wealth of Powell material, primarily in three sections: Alphabetical, Chronological, and Geographical. Marston collected articles, clippings, photocopies, microfilm, and transcripts related to Powell, his associates, his surveys and Colorado River expeditions, and his critics and apologists. There is a lively series of letters between Marston, who was decidedly critical of Powell, and William Darrah, Powell's admiring biographer (**#609**). The collection also contains photographic prints by Beaman and Hillers.

2. Library of Congress, Manuscript Division. Washington.

Letters to, from, and about Powell are found in many manuscript collections (see **#38**, **#67**, **#71**, **#78**). In the 1970s, the National Historic Preservation Commission searched LC manuscripts to find papers of important Americans that were scattered among collections. In the *NHPC Search Sheets*, v. 33, there are 17 pages with citations to Powell items in miscellaneous collections. The "Clyde Eddy Scrapbook," a collection of contemporary newspaper clippings on Powell's Colorado

River expeditions, is held in the LC's Rare Books and Special Collections.

National Archives I. Washington.

Many of the official records pertaining to Powell's Rocky Mountain Survey, his administration of the USGS, and the Irrigation Survey are housed in the National Archives and Records Administration (NARA) buildings in Washington and College Park, MD. Important collections are as follow.

3. Adjutant General's Office 1780s–1917, RG 94.

Compiled Service Records of Volunteer Soldiers contains Powell's records for both Company H, 20th Illinois Volunteer Infantry, and Battery F, Second Illinois Light Artillery. Additional details of Powell's military service are in Records of Orders, Muster Rolls, and Returns. Regimental records and histories are in Records Relating to Volunteer Organizations. Microfilm available.

4. U.S. Bureau of Indian Affairs, RG 75.

Letters Received by the Office of Indian Affairs 1824–1880 contains letters written by and about Powell during his early years of fieldwork while conducting the Powell Survey. See records for the Nevada Superintendency, 1872–1873 and 1874–1875, and the Utah Superintendency, 1873–1874 and 1875–1877, which includes the manuscript of the report of Special Commissioners Powell and Ingalls (see **#236**) and a letter of appointment by Delaney. Microfilm and accompanying pamphlet M234.

5. U.S. Veterans Administration, RG 15. Pension Office.

Records of Powell's Civil War service are in Pension File Records. His file contains letters and service records to support Powell's invalid claim, wife Emma Dean's application for widow's pension, and the Powells' marriage certificate. Powell's deposition for Walter Henry Powell's pension hearing is in Pension Case Hearings files.

National Archives II. College Park, MD.

In addition to textual material, the College Park facility houses most of the NARA's photographic and cartographic material (see **#994**, **#998**). Important collections are as follow.

6. U.S. Dept. of the Interior, Records of the Office of the Secretary of the Interior, RG 48.

The surveys of Powell and Hayden reside in the Dept. of the Interior, and the USGS is a bureau of that department. The Miscellaneous and Patents Division Series contains records relating to the USGS and the predecessor surveys, 1868–1880. Letters Received Concerning the Geological Survey includes correspondence from Powell to Secretary Carl Schurz regarding administration of the Geographical and Geological Survey of the Rocky Mountain Region and the House Committee on Public Lands. Records of the Lands and Railroads Division contains letters from a variety of constituencies regarding the closing of public lands during the time of Powell's Irrigation Survey.

7. U.S. Geological Survey, Records, RG 57.

This key collection contains the major records of the predecessor surveys of Hayden, King, Wheeler, and Powell. For additional records concerning the administration of the Powell Survey, see RG 48, Records of the Office of the Secretary of the Interior (**#6**) and Records of the Bureau of American Anthropology (**#10**, **#11**). Also see **#994** and **#998** for Cartographic and Photographic Records. A comprehensive inventory of this record group is available on a searchable CD-ROM, on site use only. See especially:

(a) U.S. Geographical and Geological Survey of the Rocky Mountain Region (Powell Survey), 1869–1882, has two essential collections for anyone researching Powell: (1) Letters Received 1869–1879 contains letters from scientists, members of Congress, secretaries of the Smithsonian and Interior, Indian agents, and more; microfilm and accompanying pamphlet M156. (2) Letters Sent 1876–1879 contains copybooks of letters to Congress, the Smithsonian, and others concerning ethnography and the establishment of the USGS; not microfilmed (some letters are illegible).

(b) Office of the Director of the U.S. Geological Survey 1839–1997, has three important accessions: (1) Letters Sent 1879–1895 contains handwritten copies of reports to the Secretary of the Interior and letters to Congress, USGS field teams, publishers, etc.; also includes typescript index to Letters Sent 1879–1883; microfilm and accompanying pamphlet M152. (2) Letters Received 1879–1901 has communications from survey employees, government officials, and citizens, and typescript index to Letters Received 1880–1889; microfilm and accompanying pamphlet M590. (3) Registers of Letters Received 1879–1901 includes notations relating to action taken with regard to incoming letters, which are significant for 1895–1901 because copies of outgoing communications are unavailable; microfilm M157.

(c) Water Resources Division, Records of the Irrigation Survey and Hydrographic Division, 1889–1890. Records Concerning Field Operations 1889–1890 contains correspondence, reports, telegrams, graphs, and charts concerning the survey in the West. Of particular significance is Powell's typewritten report outlining his position on irrigation and analyzing events during the period 1867 to 1890 that influenced the development of U.S. water policy.

New York Public Library. New York City, NY.

8. New York Public Library, Robert Brewster Stanton Papers. New York City.

This collection contains Stanton's manuscripts and extensive background material on the Colorado River, much of which bears directly on Powell. The 1,000-page manuscript, titled "The Colorado River of the West, and the Exploration, Navigation, and Survey of its Canyons, from the Standpoint of an Engineer," was later published in two volumes, *Down the Colorado* (**#164**) and *Colorado River Controversies* (**#774**). There is extensive correspondence between Stanton and Jack Sumner from 1904 through 1907 regarding Sumner's journal; typewritten accounts of interviews with Sumner (**#137**) and Billy Hawkins (**#123**); and letters from Powell and typescripts of his original 1869 journal. Also Sumner's "Story of the Powell Exploration of 1869" written in 1903 and a copy of Sumner's diary (**#56**). Portions available on microfilm. Finding aid in NYPL online catalog.

Smithsonian Institution. National Anthropological Archives. Suitland, MD.

The NAA's collections of manuscripts, correspondence, photographs, maps, and audiovisual material include the research reports and records of the BAE. Powell's diaries and field notes are among its earliest ethnographic collections. *Guide to the National Anthropological Archives* (**#966**) is a basic reference source for information on holdings. Most of the collections are cataloged in SIRIS <www.siris.si.edu>, the Smithsonian's online library and archives catalog.

9. Numbered Manuscripts, Records, and Papers.

Among the documents in this miscellany by many authors is the most extensive manuscript collection of Powell. Finding aids include the original card catalog, published as *Catalog of Numbered Manuscripts at the National Anthropological Archives* (**#962**) and SIRIS. Important manuscripts in this collection include:

(a) John Wesley Powell. Journals 1869–1872 (ms 1795a). Manuscripts and typed transcripts by William Culp Darrah. For published transcripts, see **#31–#33**, **#35**, **#100**, and **#746**.

No. 1: Journal of the Canyon trip of 1869.
No. 2: Geological notes and sections from 1869.
No. 3: Stewart's notes and sections (technical data) from the 1871–1872 expedition.
No. 5: Journal from Green River to Brown's Park, 1871–1872 expedition.
No. 8: Land trip of 1870.
No. 9: Geological notes, 1871 Kanab and 1872.
No. 10: Journal of the 1872 trip through the Grand Canyon.
Nos. 4, 6, and 7 are lost.

(b) Ethnographic manuscripts by Powell. Powell conducted linguistic, ethnographic, and folklore studies among the Indians of the Great Basin and Colorado Plateau. Between 1868 and 1880, data were recorded in 76 manuscripts. Many of these are compiled, edited, and annotated by Don D. and Catherine S. Fowler in *Anthropology of the Numa* (**#29**).

(c) Letterbook 1897–1902 (ms 3322). Indexed by name, with over 450 letters of both a personal and business nature.

(d) Typed undated manuscript (ms 4024e) of a lengthy statement by Powell refuting charges by Senator Stewart against his administration of the USGS.

10. Bureau of American Ethnology, Records, 1879–1908, RU NAA1.

General Outgoing Correspondence 1879–1907 consists of bound letterpress copybooks of Powell correspondence and includes some personal letters and material relating to the USGS.

11. Bureau of American Ethnology, Records, 1878–1965, RU NAA2.

Incoming Letters 1878–1906 includes items related to the ethnological work of the Geographical and Geological Survey of the Rocky Mountains and Powell's collection of data on Indian languages. General Correspondence 1909–1950 has records of the 1903 investigation of the administration of the BAE, including letters by Powell and information about Powell's organization of the BAE and his assumption of the director's title.

12. U.S. National Museum. Dept. of Anthropology/National Museum of Natural History. Records and Manuscripts 1897–1990 (with earlier material), RU NAA5.

> From 1879 to 1965, anthropology at the Smithsonian was divided between the Dept. of Anthropology and the BAE. This record contains some Powell material in the Division of Ethnology manuscript and pamphlet file.

Smithsonian Institution Archives. Washington.

Official records provide a history of the Smithsonian's role in 19th-century American science and in particular document the relationship among the Smithsonian, BAE, and various surveys, including the USGS under Powell's leadership. The John Wesley Powell Research File contains a name index to Powell items, predominantly correspondence in a variety of record units. *Guide to the Smithsonian Archives 1996* (**#969**) is a thorough descriptive guide to the official records of the Smithsonian.

13. Records of the Secretary's Office, 1846–1985.

> Most of the Secretary's records are available on microfilm. Corresponding letters to and from Powell may be found in records of the BAE (**#10**). See especially Office of the Secretary (Joseph Henry, Spencer F. Baird) 1863–1879, Incoming Correspondence, RU 26; Office of the Secretary (Joseph Henry, Spencer F. Baird, Samuel P. Langley) 1865–1891, Outgoing Correspondence, RU 33; and Office of the Secretary (Samuel P. Langley) 1897–1907, Outgoing Correspondence, RU 34.

14. Marcus Benjamin Papers, 1886–1929, RU 7085.

> Biographical information on Powell is included among papers gathered by Benjamin for a sketch of Powell published in *Appleton's Cyclopaedia of American Biography* (**#691**). Inexplicably, among obituary clippings and copies of various Powell documents, are vouchers and copies of invoices for expenditures for photography of the 1871–1872 exploration. There is a lengthy letter from G. K. Gilbert to Benjamin, dated 1888, which details Powell's organizational work in Washington and principal publications.

15. U.S. National Museum Administrative Records.

> USNM records include correspondence to and from Powell and documentation of material collections donated to the museum by Powell and his field assistants. See especially Assistant Secretary (Spencer F. Baird) 1850–1877, Incoming Correspondence, RU52, and

Outgoing correspondence, RU 53; Assistant Secretary in charge of the USNM 1869–1908, Incoming Correspondence, RU 189, and Letters Received from Departments and Bureaus, RU 201; USNM 1877–1975, Permanent Administrative Files, RU 192; and Office of the Registrar 1834–1958, Accession Records, RU 305.

16. Smithsonian Institution. National Museum of Natural History, Anthropology Library. Washington.

In addition to the historical collection of books originally collected by Powell and James Pilling for the library of the BAE, the library has a 10-volume bound set, *The Published Writings of John Wesley Powell.* The publications are interspersed with typewritten title pages, tables of contents, and indexes. On the title page of v. 1 is a note: "Collected and arranged for G. K. Gilbert." The text on title pages correspond exactly to the wording in P. C. Warman's bibliography of Powell, published in *Open Court* 1903 (**#980**). Two articles in the set are rare: in v. 6, "The Water Supply of the Great Plains and Its Availability for Irrigation Purposes" (**#465**) and in v. 9, "The Water and the Method" (**#463**).

17. University of Arizona Library, Special Collections, Papers of Frederick Samuel Dellenbaugh. Tucson.

This wide-ranging collection includes correspondence with and about Powell, publications by Powell, a copy of a contract for members of the 1869 expedition, and a letter of appointment to the 1872 expedition (**#40**–**#41**).

18. Utah State Historical Society Archives, William Culp Darrah Papers, ca. 1946–1972. Salt Lake City.

Manuscripts, correspondence, research notes, photographs, and other material gathered by Darrah for his biography of Powell and related work for the *UHQ*. Important material includes correspondence with descendents of Powell's siblings; a lengthy memoir by Eric Parson regarding Powell's last years in Haven, Maine; and many photocopies and transcripts of primary sources from other libraries and archives, such as a typescript of Jacob Hamblin's journals. While there are no original Powell materials, Darrah had access to and made copies of the Lyle H. Durley diary, the Rhodes Allen diary, and three letters by Andy Hall, all from the 1869 expedition. See **#36**, **#43**, and **#45**.

2. Other Collections

19. American Philosophical Society, Wallace Stegner Papers. Philadelphia.

Copies of research materials collected by Stegner for *Beyond the Hundredth Meridian* (**#671**). Stegner was a Research Fellow at the APS in 1955. The collection relates to the Powell Surveys, the Hayden-Powell controversy, and the Cope-Marsh controversy, and includes copies of letters from collections in the NARA and a transcript of William Byers' diary (**#39**).

20. Grand Canyon Museum, John Wesley Powell Collection, 1871–1947. Grand Canyon National Park, AZ.

Many transcripts and copies, pamphlets, newspaper clippings, photographs, and memorabilia related to Powell and the centennial celebration of the 1869 expedition. Includes the journals of Walter Clement Powell (**#53**) and a poem written by J. W. Powell for his mother. Not seen.

21. Illinois Wesleyan University, The Ames Library, Tate Archives and Special Collections, John Wesley Powell Collection. Bloomington.

Records of Powell's professorship at IWU include catalogs from 1865–1868, a letter from Milton Titterington recalling his participation in the Colorado expedition in 1867, and minutes of the Board of Trustees with records of honorary degrees conferred on Powell.

22. New York Public Library, Century Company Letters. New York City.

Letters from Powell related to publication of his articles and others in *Century Illustrated* magazine, including letters regarding Powell's series of articles on irrigation.

23. Newberry Library, Edward E. Ayer Manuscript Collection, Elmo Scott Watson Papers, 1816–1951. Chicago.

Watson corresponded with William C. Darrah, relatives of IWU students and faculty who accompanied Powell to Colorado in 1867 and 1868, and others to gather material for *The Professor Goes West* (**#144**) and *The Illinois Wesleyan Story* (**#783**). Contains manuscripts and resources for Watson's publications.

24. Northern Arizona University, Cline Library Special Collections and Archives, Martin J. Anderson Collection, 1809–1995. Flagstaff.

> Manuscripts, correspondence, clippings, and photocopies pertaining to Anderson's research on the 1869 and 1872 expedition. See **#1013**.

25. Shiloh National Military Park, Miscellaneous Collection, National Park Service. Shiloh National Military Park, TN.

> This collection has the original seven-page letter written by Powell to Colonel Cornelieus Cadle on 15 May 1896, in which he recreates his participation in the Battle of Shiloh, culminating in the injury to his right arm that necessitated amputation. Not seen.

26. Southern Illinois University, Morris Library, Open Court Publishing Company Records. Carbondale.

> Open Court published several articles by Powell in its journals *The Monist* and *Open Court*, as well as Powell's philosophical treatise *Truth and Error* (**#510**). The majority of the 20 letters from Powell, dated 1898 to 1900, had to do with deadlines, galleys, and other business. Several personal letters show a developing warm relationship between Powell and Paul Carus, Open Court's editor.

27. Utah State Historical Society, John Wesley Powell Collection, 1947–1950. Salt Lake City.

> Correspondence and drafts for articles in *UHQ* 15 (**#102**) and 16–17 (**#103**) on the Powell expeditions.

28. Western Reserve Historical Society Library, Joseph Stanley-Brown Papers, 1874–ca. 1930. In Mary Garfield Stanley-Brown Papers. Cleveland, OH.

> Joseph Stanley-Brown served briefly as Powell's stenographer during 1879, participated in Geological Survey fieldwork (1888–1890s), and married James Garfield's daughter. In his unpublished autobiography, "Reminiscences," Stanley-Brown refers to his supporting role in the consolidation of the surveys in 1879, "accompanying [Powell] to newspaper offices to secure wide distribution of propaganda literature and to the [C]apitol to importune Members and Senators to support the measure…." The manuscript of his eulogy to Powell, read before the Washington Literary Society on 13 Dec. 1902, is an uncritical and passionate tribute to a friend and acquaintance, notable for Stanley-Brown's personal observations of Powell's demeanor during his testimony before Congress in 1879.

B. PUBLISHED MANUSCRIPTS OF JOHN WESLEY POWELL

29. *Anthropology of the Numa: John Wesley Powell's Manuscripts on the Numic Peoples of Western America, 1868–1880*. Ed. by Don D. and Catherine S. Fowler. Smithsonian Contributions to Anthropology, Vol. 14. Washington: Smithsonian Institution Pr, 1971.

> Between 1868 and 1880, Powell conducted linguistic, ethnographic, and folklore studies among the Numic-speaking Indians of the Great Basin and Colorado Plateau. The data were recorded in over 70 unpublished manuscripts, now in the NAA (**#9**), which are collated, edited, and annotated in this volume. Materials include general ethnographic data, a variety of myths and tales, and extensive vocabulary lists of various Numic languages and dialects. In addition, "Report of J. W. Powell and G. W. Ingalls" is reprinted (**#236**). The extensive introduction contains a brief biographical sketch of Powell, a description of his ethnographic fieldwork and itinerary for 1868–1880, a description of the Indian groups with which Powell worked, and an assessment of his fieldwork. The manuscripts section contains a discussion of the manuscripts, followed by transcriptions of manuscripts organized into linguistic groupings. Two of the manuscripts are papers read before the Philosophical Society of Washington in 1874: "On the Genesis and Demonology of the Numa Tribe of Indians" and "Mythology of the Numas." Includes an excellent bibliography with references to Powell's published works on anthropology.

30. "Extracts From John Wesley Powell's Notes on the Indians." *John Wesley Powell and the Anthropology of the Canyon Country*. Ed. Don D. Fowler, Robert C. Euler, and Catherine S. Fowler. U.S. Geological Survey Professional Paper 670. Washington: U.S. Dept. of the Interior, Geological Survey, 1969. 22–28.

> Transcript of material on the Ute people from Powell's collection of manuscripts (see **#9**). Topics are "Means of Subsistence," "Animal Food," "Courtship and Marriage," and "Mythology and Beliefs."

31. "Geological Notes and Sections by J. W. Powell." Ed. and transcr. William Culp Darrah. *UHQ* 15 (1947): 134–139.

> Powell's second journal from the 1869 expedition contains geological notes and sections, and various tables of data mostly of an astronomical nature. The journal corresponds with the first notebook containing the actual diary of the second leg of the exploration. See **#9** and **#735**.

32. "J. W. Powell's Astronomical Observations 1869." *First Through Grand Canyon*. Ed. Michael P. Ghiglieri. Transcr. Richard Quartaroli. Flagstaff, AZ: Puma Pr, 2003. Appendix, 302–313.

Powell's astronomical notes from 1869, in Journal 2 at the NAA (**#9**).

33. "John Wesley Powell's Journal: Colorado River Exploration 1871–1872." Ed. and transcr. Catherine S. and Don D. Fowler. *Smithsonian Journal of History* 3.2 (1968): 1–44. [Rpt. *Canyon Legacy* 5 (1990): 2–12.]

Portions of the original journals kept by Powell while on the river during the 1871 and 1872 trips. Ethnographic material is omitted. See NAA (**#9**).

34. [Letters]. *Stephen Powers, California's First Ethnologist; and, Letters of Stephen Powers to John Wesley Powell Concerning* Tribes of California. By Susan Park. Contributions of the University of California Archaeological Research Facility 28. Berkeley: Archaeological Research Facility, Dept. of Anthropology, U of California, 1975.

This book consists of Part. 1, "The Life of Stephen Powers," by Susan Park (**#150**), and Part. 2, "Letters of Stephen Powers to John Wesley Powell Concerning *Tribes of California*," edited by R. F. Heizer. Part. 2 includes three letters from Powell to Powers written between 1880 and 1882.

35. "Major Powell's Journal, July 2–August 28, 1869." Ed. and transcr. William Culp Darrah. *UHQ* 15 (1947): 125–131.

Journal 1 in the NAA (**#9**). See also **#33**, **#100**, and **#746**.

C. MANUSCRIPT AND ARCHIVAL SOURCES FOR ASSOCIATES OF JOHN WESLEY POWELL

This category includes important associates who influenced him or whose papers contain correspondence with Powell, offer comment on him, or record activities carried on in his behalf.

1. Expeditions and Surveys

36. Allen, Rhodes. Journal. In William C. Darrah Papers. Utah State Historical Society. Salt Lake City.

Typescript of an IWU student's diary from the 1868 Colorado expedition

37. Bishop, Francis Marion. Papers. Utah State Historical Society. Salt Lake City.

Powell was one of Bishop's professors at IWU and a member of the 1867 Colorado expedition. He served as topographer of the 1871–1872 expedition and made the first map of the lower Colorado River. In addition to two journals from the later expedition, the collection includes transcripts of letters written to the Bloomington *Pantagraph* and a transcript of a lecture delivered in 1875 at Salt Lake City (published in *UHQ* 15, **#90**).

38. Bradley, George Young. Papers. Library of Congress, Manuscript Division. Washington.

Bradley was one of Powell's chief boatmen and assistant in the geological work. This is the most complete and literate diary of the 1869 expedition. In two parts, 21 May–28 June and 29 Jun–30 Aug. Published in *UHQ* 15 (**#97**); see **#100** and **#746**.

39. Byers, William Newton. Diary. Denver Public Library, Western History and Genealogy. Denver, CO.

Byers, editor of the *Rocky Mountain News*, briefly joined Powell's 1868 expedition. The diary and published articles in the *News* record his experience (see also article in *The Trail*, **#98**). Denver Public Library also has an index to Western periodicals and newspapers and an extensive collection of photographs from the 19th-century surveys (see **#1020**).

40. Dellenbaugh, Frederick Samuel. Papers. New York Public Library.

Artist and boatman Dellenbaugh wrote the only published account of the 1871–1872 expedition. This collection contains extensive notes, letters, and manuscripts, including his diary, 12 Apr. 1872–4 Apr. 1873. He used his experience in *The Romance of the Colorado River* (**#109**), but a fuller and more personal account was published as *A Canyon Voyage* (**#104**).

41. ——. Papers. University of Arizona Library, Special Collections. Tucson.

This collection contains notes for *A Canyon Voyage* (**#104**). Correspondence includes letters from S. V. Jones, A. H. Thompson, Jack Sumner, and William Bass (many of which discuss Powell's character). See also **#17**.

42. ——. Papers, ca. 1926–1935. Utah State Historical Society. Salt Lake City.

Primarily correspondence from Dellenbaugh to Charles Kelly and others during the 1920s and 1930s regarding the Colorado River and his service with Powell's second Colorado River expedition. See *UHQ* 37 (**#110**).

43. Durley, Lyle. Journal. In William C. Darrah Papers. Utah State Historical Society. Salt Lake City.

Typescript of an IWU student's diary from the 1868 Colorado expedition.

44. Garman, Samuel. Papers. Illinois State University Archives. Normal.

Garman was a student who accompanied Powell's exploring expedition to the Rocky Mountains in 1868. There are four letters from Colorado and Nevada in 1868 and 1869 written to Gertrude Lewis. He describes the ascent of Long's Peak and hints at conflicts with Emma Powell.

45. Hall, Andrew. Letters. In William Culp Darrah Papers. Utah State Historical Society. Salt Lake City.

The youngest member of the 1869 expedition sent letters to his mother and brother. The collection contains copies of three letters and a photograph. One original letter is in the Grand Canyon Museum, the others remain with family. Published in *UHQ* 16–17 (**#120**).

46. Hawkins, William W. "William Hawkins' Story." In Robert Brewster Stanton Papers. New York Public Library.

"Billy" Hawkins did not keep a diary of his experience with the 1869 expedition, but prepared this narrative for Stanton in 1907, later published in *Colorado River Controversies* (**#123**). See also **#100**, **#122**, and **#746**.

47. Hayden, Ferdinand Vandiveer. Records of the Geological and Geographical Survey of the Territories, 1847–1881, RG 57. National Archives II. College Park, MD. Microfilm M623.

This record group contains the bulk of the official records of the Hayden Survey. There are smaller amounts of material in RG 48 (**#6**) and 49.

48. Hillers, John K. Diaries. In Numbered Manuscripts, Records, and Papers. National Anthropological Archives. Suitland, MD.

Hillers signed on as a boatman for the second Grand Canyon expedition. He stepped in as its photographer and later became well known as a USGS photographer. The two journals from 1871–1872 and one from 1875 were published in *Photographed All the Best Scenery* (**#125**).

49. Jones, Stephen Vandiver. Papers. New York Public Library.

Jones was Thompson's assistant topographer on the second Grand Canyon expedition. There are two small diaries from 1871 and 1872; a journal of daily measurements and technical notes from 1872; and miscellaneous papers. The journals were published in *UHQ* 16–17 (**#127**).

50. Keplinger, Lewis Walter. Papers. Kansas State Historical Society. Topeka.

Keplinger was Powell's student at IWU. This collection includes letters written to his brother during the 1868 expedition to the Rocky Mountains. There is also a letter from Jack Sumner, written in 1906, accusing Powell of abusing Dunn and the Howland brothers. Microfilm available. See also **#98** and **#139**.

51. King, Clarence. Papers, 1859–1902. In Papers of James D. Hague. Henry E. Huntington Library. San Marino, CA.

King was the first director of the USGS. Personal and business papers relating to the work of the California Survey and King's Fortieth Parallel Survey, assembled by Hague. See also **#52** and **#73**.

52. ——. Records of the United States Geological Exploration of the Fortieth Parallel (King Survey), 1867–1881, RG 57. National Archives II. College Park, MD.

King led one of the four "Great Surveys" and was the first director of the USGS (1879–1881). Microfilm M622. For records of King's administration, see **#51** and **#73**.

53. Powell, Walter Clement. Journals. Grand Canyon Museum. Grand Canyon National Park, AZ.

"Clem" was Powell's cousin. He learned photography for the second Grand Canyon expedition and become a permanent member of Powell's staff. There were originally four notebooks, but the third was lost. Published in *UHQ* 16–17 (**#133**); see **#746**.

54. Steward, John F. Journal, May 22–November 3, 1871. American Geographical Society. New York City.

Steward served as assistant geologist on the second Powell Colorado River expedition. In 1903, Steward sent Dellenbaugh a typescript of his diary (the original was written in shorthand), which Dellenbaugh gave to the American Geographical Society in 1909. The NYPL has a copy annotated by Dellenbaugh (**#40**). The journal was published in *UHQ* 16–17 (**#136**). Also in this collection are letters from Dellenbaugh and a letter by Steward written from the river and addressed to an unknown editor, dated July 1871. Not seen.

55. ——. Field Notes. In Numbered Manuscripts, Records, and Papers. National Anthropological Archives. Suitland, MD.

Steward's field notes are preserved in ms 1795a, Powell's Notebook no. 3, "Steward's Notes and Sections" (**#9**). The geological sections measured by Steward are given in full in the *Report on the Geology of the Eastern Portion of the Uintah Mountains* (**#249**).

56. Sumner, John Colton. Journal, May 24–June 28, 1869. In Numbered Manuscripts, Records, and Papers, ms 4419. National Anthropological Archives. Suitland, MD.

Jack Sumner accompanied Powell on the 1869 expedition and became a bitter protagonist in following decades. This portion of his journal was mailed by Powell from the Uinta Indian Agency on 6 July to the St. Louis *Missouri Democrat*. It was published 24–26 Aug. under the caption "From Col. Powell ... Journal of Jack Summers [*sic*]." In 1911, Stanton found several loose sheets of Sumner's original diary in his hand, dated 17–27 Aug. interleaved with Powell's 1869 journal. See **#57** for the second half of Sumner's journal. Published in *UHQ* 37 (**#140**); see **#100** and **#746**.

57. ——. Journal Transcript, July 6–August 31, 1869. In Robert Brewster Stanton Papers. New York Public Library.

In 1907, Stanton obtained what was believed to be the original diary kept by Sumner. The journal is most likely a copy made by someone at the Smithsonian from a copy that Sumner himself had made for Powell shortly after the completion of the expedition. The copy now preserved in the Stanton collection (**#8**) was obtained from the Smithsonian in 1907. It is a copy of the NYPL copy, along with the field notes in Sumner's handwriting, now in the permanent collection of the Smithsonian. The Stanton collection also contains extensive correspondence with Sumner 1904–1907 over this journal and his narrative of 1869 as told to Stanton (**#137**). Published in *UHQ* 15 (**#138**); see **#100** and **#746**. For the first half of the journal, see **#56**.

58. Thompson, Almon Harris. Journals. New York Public Library.

Thompson was Powell's brother-in-law, topographer for the second Powell expedition, and Chief Geographer of USGS. There are five diaries from 1871 to 1875. The first three parts—navigation of the Green-Colorado in 1871–1872, exploration from Kanab to the Fremont River, and mapping of Utah and Arizona north of the Grand Canyon—are published in *UHQ* 7 (**#142**). The other two are published in expanded form in *Exploration*.

59. Thompson, Ellen Powell. Diary, 1872–1873. In Almon Harris Thompson Collection. New York Public Library.

The wife of A. H. Thompson accompanied her husband through Utah and Arizona in 1871–1872. Her diary is kept with Thompson's journals. Her plant collection is preserved in the Gray Herbarium at Harvard University. The diary is published in *UHQ* 62 (**#143**).

60. Wheeler, George Montague. U.S. Geographical Surveys West of the One Hundredth Meridian (Wheeler Survey), 1871–1880. Records of the Office of the Chief of Engineers RG 77. National Archives II. College Park, MD.

Most of the textual, cartographic, and photographic records of the Wheeler Survey are in this group. RG 57 also has a small collection of Wheeler papers (see #7).

61. Wood, Henry Ellsworth. Papers. Henry E. Huntington Library. San Marino, CA.

At the age of 13, Henry Wood accompanied his father, Rev. William Cowper Wood, to Colorado with Powell in 1968. The collection includes his father's notebook from that expedition, "Colorado in 1868," filled with lively accounts of life on the frontier with Emma and John Powell.

2. Bureau of American Ethnology

62. Baird, Spencer F. Papers, 1833–1889, RU 7002. Smithsonian Institution Archives. Washington.

Second Secretary of the Smithsonian Institution, 1878–1887. Incoming and Outgoing Correspondence includes correspondence with Powell. Microfilmed. See also records of the USNM, RU 52 and 53 (**#15**).

63. Boas, Franz. Papers. American Philosophical Society, Philadelphia.

Boas was a central figure in American anthropology. He edited the *Handbook of American Indian Languages* (**#798**). There are six letters from Powell in correspondence from 1887–1895 regarding Boas's work with Indian languages and an offer for the position of editor of BAE publications. The collection is available on microfilm as Professional Correspondence, 1862–1942.

64. Holmes, William Henry. Papers, 1871–1931, RU 7084. Smithsonian Institution Archives. Washington.

Scientific illustrator, geologist, ethnologist, curator of the National Museum, and head of BAE after Powell. The collection includes correspondence from Powell.

65. ——. "Random Records of a Lifetime, 1846–1931: Cullings, Largely Personal, From the Scrap Heap of Three Score Years and Ten, Devoted to Science, Literature and Art." Smithsonian Institution, Library of the National Museum of American Art and the National Portrait Gallery. Washington.

Typewritten manuscript and material for an unpublished autobiography, bound in 20 volumes (vols. 17–20 are in the possession of the family). Of note: v. 5, With Powell and Langley in the Rocky Mountains, 1887; v. 8, With Langley and Powell in Cuba, 1900; v. 10, Powell's Death and Powell Memorial; and v. 14, Portrait and Biography of Powell. Several letters from Powell are scattered throughout. Microfilm (7 reels) and finding aid available in the library.

66. Langley, Samuel P. Papers, 1867–1906, RU 7003. Smithsonian Institution Archives. Washington.

Third Secretary of the Smithsonian Institution, 1887–1906. Langley accompanied Powell to Cuba in 1900.

67. McGee, WJ [William John]. Papers. Library of Congress, Manuscript Division. Washington.

McGee, who insisted on "WJ" as his name, was Powell's protégé, first as a geologist with the USGS, then as ethnologist for the BAE. See **#68**.

68. ——. Bureau of American Ethnology, Records, 1879–1965, RU NAA1. National Anthropological Archives. Suitland, MD.

Outgoing Letters 1893–1903 contains letters to Powell corresponding to Incoming Letters in the Library of Congress (**#67**).

69. Morgan, Lewis Henry. Papers. University of Rochester, Rhees Library. Rochester, NY.

Morgan played a significant role in the development of Powell's theories and work in anthropology. This collection includes a series of letters from Powell. Not seen.

70. Pilling, James Constantine. Papers, RU NAA2. National Anthropological Archives. Suitland, MD.

Pilling began compiling extensive bibliographies while clerk for the Rocky Mountain Survey. He was chief clerk of the BAE as well as the USGS. For a complete listing of Pilling's bibliographies, see BAE Bulletin 49 (**#972**).

3. U.S. Geological Survey

71. Becker, George. Papers. Library of Congress, Manuscript Division. Washington.

Becker was a friend and appointee of Clarence King, continuing service during Powell's administration. There are over 70 letters, mostly routine, from Becker to Powell, 1887–1889.

72. Dall, William H. Papers, RU 7073. Smithsonian Institution Archives. Washington.

Dall was paleontologist with the USGS and honorary curator of the USNM. Correspondence with Powell regarding collections.

73. King, Clarence. Records of the Office of the Director of the U.S. Geological Survey, RG 57. National Archives II. College Park, MD.

Official correspondence is in Letters Sent 1879–1895. See **#51–#52**.

74. Marsh, Othniel Charles. Papers, 1817–1899. Yale University, Beinecke Rare Books and Manuscripts Library. New Haven, CT.

Yale professor appointed to head the paleontology division of the USGS and Powell's ally in battles with Congress and Edward Cope. The collection has some letters from Powell and papers related to the National Academy of Science. Not seen.

75. Ward, Lester Frank. Papers. Brown University, John Hay Library. Providence, RI.

Ward was a close associate of Powell and a paleobotanist for the USGS, 1881–1907. He was a pioneer of modern and evolutionary sociology whose most famous work is *Dynamic Sociology* (**#152**). Some correspondence with Powell. Available on microfilm. Not seen.

4. Other

76. Agassiz, Alexander. Papers. Harvard University, Ernst Mayr Library. Cambridge, MA.

A prominent Harvard professor and scientist opposed to a role for government in science. There are three letters from Powell to Agassiz and a letter from Dutton to Agassiz (24 Apr. 1878) regarding charges against Powell.

77. Anthropological Society of Washington. Numbered Manuscripts, Records, and Papers, ms 4821. National Anthropological Archives. Suitland, MD.

Powell was a founding member of the ASW in 1879. This collection contains the original minutes and abstracts, published as *Abstract of Transactions of the ASW* (**#288**).

78. Bell, Alexander Graham. Alexander Graham Bell Family Papers. Library of Congress, Manuscript Division. Washington.

The inventor of the telephone had an interest in a wide range of scientific and technological fields. This large collection includes correspondence with Powell, Ulysses S. Grant, Joseph Henry, S. P. Langley, and Simon Newcomb. Includes letter from Emma Dean Powell seeking Bell's assistance in securing a pension in 1902. Most letters to or about Powell have been digitized and are available online at the LC's *American Memory* website <memory.loc.gov/ammem/amhome.html> (**#1011**).

79. Bleak, James Godson. Papers, 1864–1895. Utah State Historical Society. Salt Lake City.

Papers of this Mormon Church historian include copies of a manuscript, "Historical Memoranda of Southern Utah," and a typescript of *Annals of the Southern Utah Mission*, which has copies of letters and other documentary material relating to Powell and Jacob Hamblin in the early 1870s.

80. Garfield, James A. Papers. Library of Congress, Manuscript Division. Washington.

> As both U.S. representative and president, Garfield supported Powell's work in the 1870s. These papers contain a highly critical report on Hayden by Powell, dated 7 Mar. 1879. Microfilm available.

81. Grant, Ulysses S. Papers. Library of Congress, Manuscript Division. Washington.

> Powell and Grant's relationship extended from the Civil War through Grant's presidency. Microfilm available.

82. Henry, Joseph. Collection, 1808–1878, RU 7001. Smithsonian Institution Archives. Washington.

> First Secretary of the Smithsonian, 1846–1878.

83. Merrill, George P. Collection, ca. 1800–1930, RU 7177. Smithsonian Institution Archives. Washington.

> A collection of photographs and autographs, Powell's among them, for Merrill's history of American geology (**#856**).

84. Philosophical Society of Washington, 1871–1968, RU 7079. Smithsonian Institution Archives. Washington.

> Powell was an active member and officer of the PSW. The collection contains minutes of meetings, copies of lectures, and PSW publications.

85. Stewart, William Morris. Papers. Nevada Historical Society. Reno.

> Letterbooks 1888–1891. Letters to various individuals reveal Senator Stewart's grievances against Powell. Not seen.

86. Washington Academy of Sciences, 1888–1972, RU 7099. Smithsonian Institution Archives. Washington.

> Powell was an officer of the Academy. This collection contains minutes and scrapbooks.

D. PUBLISHED PAPERS, DOCUMENTS, AND MEMOIRS OF ASSOCIATES OF JOHN WESLEY POWELL

1. Expeditions and Surveys

87. Bass, William Wallace, et al. *Adventures in the Canyons of the Colorado.* Foreword by George Wharton James. Grand Canyon, AZ: The Authors, 1920.

> Bass explains that he wants to tell the truth about the murder of Dunn and the Howlands because Powell is unreliable. He presents a manuscript by William Hawkins, written just before he died in 1919 (**#122**) and James White's letter of 1867 (**#147**). See also **#100**, **#123**, and **#746**.

88. Beaman, E. O. "The Cañon of the Colorado, and the Moquis Pueblos: A Wild Boat-Ride Through the Cañons and Rapids—a Visit to the Seven Cities of the Desert—Glimpses of Mormon Life." *Appleton's Journal* 11 (1874): [480]–484, [513]–516, [545]–548, 590–593, 623–626, [641]–644, 686–689.

> Beaman, photographer for the Powell Survey, beat Powell to press with this seven-part detailed narrative of the second Colorado River expedition. Extracts from his journal present details of daily life on the river in early 1871 through his departure at Kanab. The rest of the series gives an account of his subsequent travel overland to the Grand Canyon and through Navajo and Hopi country.

89. ——. "The Colorado Exploring Expedition." *Anthony's Photographic Bulletin* 2 (1872): 463–465.

> The Anthony brothers recommended Beaman to Powell. In this letter from Kanab, dated 20 Dec. 1871, Beaman gives a brief account of the first leg of the second Colorado River expedition and a visit to an Indian agency.

90. Bishop, Francis Marion. "Captain Francis Marion Bishop's Journal, August 15, 1870–June 3, 1872." Ed. and transcr. Charles Kelly. *UHQ* 15 (1947): 159–237.

> For annotation, see **#37**.

91. ——. "John Wesley Powell, and His Exploration of the Colorado River." *Utah Society of Engineers Monthly Journal* 7 (1922): 217–225.

> An address to the Engineering Council of Utah in 1922. Same text as his 1918 address (**#93**).

92. ——. "Letters of Captain F. M. Bishop to the *Daily Pantagraph* 1871–72." *UHQ* 15 (1947): 239–253.

Reprinted from the Bloomington (Illinois) *Daily Pantagraph*, 7 July and 14 Sept. 1871, and 6 Feb. and 27 Mar. 1872. In addition to notes from the river, Bishop reflects and reports on encounters with Mormons and Indians. See also *Professor Goes West* (**#144**).

93. ——. "Personal Reminiscences of John W. Powell." *Transactions of the Utah Academy of Sciences* 2 (1918): 16–27.

In an address to the Academy in April 1918, Bishop presents a biography of Powell and relays his own experiences as a student with the Colorado Rockies expedition and in 1870–1871 as a member of the second Colorado River expedition.

94. Bowles, Samuel. *Our New West; Records of Travel Between the Mississippi River and the Pacific Ocean.* Hartford, CT: Hartford Pub Co, 1869.

Pages 502–503 are essentially the same as *Switzerland of America* (**#96**).

95. ——. *The Parks and Mountains of Colorado; a Summer Vacation in the Switzerland of America, 1868.* Ed. by James H. Pickering. Norman: U Oklahoma Pr, 1991.

Originally published as *The Switzerland of America* in 1869 (**#96**). Pickering's introduction and notes provide extensive background and context.

96. ——. *The Switzerland of America; a Summer Vacation in the Parks and Mountains of Colorado.* Springfield, MA: S. Bowles, 1869.

Bowles was a member of the Colfax party that undertook a much-publicized journey across the plains to the Colorado Rockies. His book contains a report on meeting with Powell and his party in Middle Park, Colorado, in August 1868 (pp. 81–86). Bowles comments on Powell's ascent of Long's Peak and becomes among the first to report that Powell, following winter in camp near the Utah border, will undertake a campaign whose "great and final object ... is to explore the upper Colorado River and solve the mysteries of its three hundred mile canyon." See **#94** and **#95**.

97. Bradley, George Young. "George Y. Bradley's Journal, May 24–August 30, 1869." Ed. William Culp Darrah. *UHQ* 15 (1947): 31–72.

For annotation, see manuscript (**#38**). Portions reprinted in **#100** and **#746**.

98. Byers, William N. "First Ascent of Long's Peak." *The Trail* 7.5 (1914): 21–23.

Byers spent several days with Powell and his party in the Rocky Mountains in August 1868. He describes their ascent of Long's Peak. Reprinted from an article in the *Rocky Mountain News*, 1 Sept. 1868. See also **#128** and **#139**.

99. Colburn, Justin E. "The Cañons of the Colorado." *Picturesque America; or, The Land We Live in.* Ed. William Cullen Bryant. Vol. 2. New York: D. Appleton and Co, 1874. 503–511.

Colburn was a reporter for *The New York Times* who accompanied Thomas Moran, Powell, and Jack Hillers to the Grand Canyon in July and August 1873. He describes the lower canyons of the Colorado River below junction of the Grand and Green rivers, exaggerating the depth of the Grand Canyon by 2200 ft. Based on his articles published in *The New York Times* in 1873: "The Great American Desert" on 15 Jul., "The Land of Mormon" on 7 Aug., and "The Colorado Cañon" on 4 Sep. Illustrations by Thomas Moran.

100. Cooley, John R. *The Great Unknown: The Journals of the Historic First Expedition Down the Colorado River*. Flagstaff, AZ: Northland, 1988.

Cooley uses direct quotes from daily journal entries and later published accounts of members of the 1869 expedition to re-create the journey from multiple perspectives. Sources include *Canyons of the Colorado*, reminiscences of Sumner and Hawkins from Bass (**#87**) and Stanton (**#137**), and various journals and letters published in *UHQ* 15 (**#102**) and 16–17 (**#103**).

101. Daniels, W. H. "Journal Leaves From Powell's Expedition of 1868." *Illinois Wesleyan University Alumni Journal* 1 (1872): 42–44.

Article signed by "Twig" (Daniels), dated 21 July 1868, recounts the ascent of the party from Denver into the Rocky Mountains. Reprinted in *Professor Goes West* (**#144**).

102. Darrah, William Culp. *The Exploration of the Colorado River in 1869; Biographical Sketches and Original Documents of the First Powell Expedition of 1869.* Spec. issue of *UHQ* 15 (1947). 1–253.

> Darrah wrote and edited the biographies and transcribed the journals, with the exception of contributions by Charles Kelly and Ralph Chamberlain on Francis Marion Bishop and his journal. With *UHQ* 16–17 (**#103**), this forms an invaluable documentary history of the Powell survey. It includes Powell's letters newspapers and his journal, letters of O. G. Howland and W. H. Powell, the second half of Jack Sumner's diary, and more. All articles are cited individually (see **#31**, **#35**, **#90**, **#92**, **#97**, **#126**, **#135**, **#138**, **#568**, **#569**, **#608**, **#718**, **#735**, **#737**, **#738**, **#882**, **#884**–**#888**).

103. Darrah, William Culp, et al. *The Exploration of the Colorado River and the High Plateaus in 1871–72. UHQ* 16–17 (1948–1949). 1–508.

> Includes journals of S. V. Jones, John Steward, and Walter C. Powell, and three letters by Andrew Hall. Biographies and transcriptions contributed by Darrah, Charles Kelly, and Herbert Gregory. With *UHQ* 15 (**#102**), an essential publication. All articles are cited individually (see **#120**, **#127**, **#133**, **#134**, **#136**, **#883**, **#888**, **#895**, **#896**).

104. Dellenbaugh, Frederick Samuel. *A Canyon Voyage: The Narrative of the Second Powell Expedition Down the Green-Colorado River From Wyoming, and the Explorations on Land, in the Years 1871 and 1872.* New York: G. P. Putnam's Sons, 1908.

> The NYPL catalog describes this publication as "Practically a second volume of *The Romance of the Colorado River*" (**#109**). Written for a popular audience, it stands as the only nearly complete, detailed, first-hand account of the second Colorado River expedition. Dellenbaugh draws on his own journal (**#40**) and those of other expedition members. The illustrations are from the author's sketches and photographs by Beaman. Republished many times. See **#105** and **#106**.

105.——. *A Canyon Voyage: The Narrative of the Second Powell Expedition Down the Green-Colorado River From Wyoming, and the Explorations on Land, in the Years 1871 and 1872.* [2nd] ed. New Haven, CT: Yale U Pr; London: H. Milford, Oxford U Pr, 1926.

> This edition has a new introduction, but in other respects is the same as the 1908 edition. See **#104**.

106.——. *A Canyon Voyage: The Narrative of the Second Powell Expedition Down the Green-Colorado River From Wyoming, and the Explorations on Land,* *in the Years 1871 and 1872.* Yale Western Americana Series. New Haven, CT: Yale U Pr, 1962. [Rpt. Tucson: U of Arizona Pr, 1984.]

Reprint of the 1926 edition (**#105**) with a new foreword by William H. Goetzmann. Provides biographical information on members of the exploring party.

107.——. "The Great Walled River." *Bulletin of the American Geographical Society* 19 (1887): 113–163.

A history of the discoveries of and descents into the Grand Canyon. Dellenbaugh supports the idea that Powell was the first to pass through the entire canyon and relates several anecdotes of the 1871–1872 expedition.

108.——. "One Night in the Grand Canyon." *Told at the Explorers Club: True Tales of Modern Exploration.* Ed. Frederick A. Blossom. New York: Albert & Charles Boni, 1931. 69–80.

The first page is a synopsis of Dellenbaugh's life, followed by his account of a rainy day in the Grand Canyon when a rapid rise in the river forced the crew to spend a night on a high, narrow ledge.

109.——. *The Romance of the Colorado River; the Story of Its Discovery in 1540, With an Account of the Later Explorations, and With Special Reference to the Voyages of Powell Through the Line of the Great Canyons.* New York: G. P. Putnam's Sons, 1902. [Rpt. Chicago: Rio Grande Pr, 1962; Alexandria, VA: Time-Life Bks, 1982; Mineola, NY: Dover, 1998.]

Dellenbaugh states that he is putting the second Colorado River expedition on record "for the first time in this volume." The first four chapters review the history of Grand Canyon exploration by Europeans and Americans. Dellenbaugh describes the men and events of 1870–1872 in detail. The book includes an epilogue, "John Wesley Powell, A Brief Review of His Career" (**#612**), in which Dellenbaugh calls attention to the fact that Powell ignored the men and work of the second expedition in his 1875 report. Illustrations are Dellenbaugh's own sketches and photographs by Hillers and Beaman. G. P. Putnam's published this again in 1903, 1906, and 1909.

110. Dellenbaugh, Frederick Samuel, and C. Gregory Crampton. "F. S. Dellenbaugh of the Colorado: Some Letters Pertaining to the Powell Voyages and the History of the Colorado River." Ed. and intro. C. Gregory Crampton. *UHQ* 37 (1969): 214–243.

The transcribed letters from the Utah State Historical Society (**#42**) were written to Russell G. Frazier, Raymond T. Stites, and Charles

Kelly. Topics include the Powell river voyages and surveys, the deaths of the Howlands and Dunn, river navigation, Mormons and the Mountain Meadows Massacre, and other historical events.

111. DeMotte, H. [Harvey] C. "From the Pahria to the Kaibab." *Illinois Wesleyan University Alumni Journal* 3 (1873): 39–43.

During this leg of the 1872 survey, Powell, Thompson, DeMotte, and others join Hillers, Jones, and Dellenbaugh on a Colorado River boat ride to the home of Mr. and Mrs. John D. Lee. Chronologically follows "Six Days on the Kaibab" (**#114**). Reprinted in *Professor Goes West* (**#144**).

112.——. "Notes by the Way." *Illinois Wesleyan University Alumni Journal* 4 (1874): 188–189.

DeMotte, a faculty colleague from IWU, arrives in Salt Lake City in July 1874 with his wife and members of the Powell expedition's "Summer campaign." In *Professor Goes West* (**#144**).

113.——. "Outward Bound." *Illinois Wesleyan University Alumni Journal* 3 (1873): 241–245.

First account of DeMotte's trip from Illinois to Salt Lake City in July 1872 to join the Powell Expedition. Chronologically precedes "Twelve Days with the Broncos" (**#116**). In *Professor Goes West* (**#144**).

114.——. "Six Days on the Kaibab." *Illinois Wesleyan University Alumni Journal* 2 (1872): 233–239.

In August 1872, Powell, Thompson, Thompson's wife, and DeMotte set off from Kanab to the Grand Canyon. On the return trek, Powell named an open meadow "DeMotte's Park." Chronologically follows "Twelve Days with the Broncos" (**#116**) and precedes "From the Pariah" (**#111**). In *Professor Goes West* (**#144**).

115.——. "Sketches Beyond the Rocky Range." *Illinois Wesleyan University Alumni Journal* 6 (1876): 241–245.

This article, with "Notes by the Way" (**#112**), contains excerpts from the journal of Mrs. DeMotte, including her observations on Salt Lake City, Mormon women, and polygamy. In *Professor Goes West* (**#144**).

116.——. "Twelve Days With the Broncos." *Illinois Wesleyan University Alumni Journal* 4 (1874): 225–228.

DeMotte's account of the first leg of his assignment with the Powell Expedition, from Salt Lake City to Kanab the summer of 1872. Chronologically follows "Outward Bound" (**#113**). In *Professor Goes West* (**#144**).

117. Dutton, Clarence E. *Report on the Geology of the High Plateaus of Utah.* Washington: GPO, 1880.

This is Dutton's first major publication on the subject of the geology of the Colorado River plateau. The prefatory note is by Powell (see **#281**). The author's preface acknowledges Powell's assistance and support, praising him for his analytic power and "new range of geological thought and philosophy." A publication of the U.S. Geographical and Geological Survey of the Rocky Mountain Region; issued with accompanying atlas, *Topographical and Geological Atlas of the District of the High Plateaus of Utah* (Washington: GPO, 1879).

118.——. *Tertiary History of the Grand Cañon District.* Monographs of the United States Geological Survey, 2. Washington: GPO, 1882. [Rpt. Santa Barbara, CA: Peregrine Smith, 1977.]

This survey of the Grand Canyon district has been deemed a classic publication in geology by virtue of Dutton's superb writing, the beautiful line drawings of panoramic views by William H. Holmes, the drawings and paintings by Thomas Moran, and the quality of its companion volume, *Atlas to Accompany the Monograph on the Tertiary History of the Grand Cañon District* (Washington: GPO, 1882).

119. Ghiglieri, Michael P. *First Through Grand Canyon; the Secret Journals and Letters of the 1869 Crew Who Explored the Green and Colorado Rivers.* Flagstaff, AZ: Puma Pr, 2003.

See **#746** for annotation and cross-references.

120. Hall, Andrew. "Appendix: Three Letters by Andrew Hall." Ed. and intro. William Culp Darrah. *UHQ* 16–17 (1948–1949): 505–508.

The youngest member of the 1869 expedition sends letters to his mother and brother.

121. Hamblin, Jacob, and James A. Little. *Jacob Hamblin; a Narrative of His Personal Experience, As a Frontiersman, Missionary to the Indians and Explorer Disclosing Interpositions of Providence, Severe Privations, Perilous*

Situations and Remarkable Escapes. Faith-Promoting Series, 5. Salt Lake City, UT: Juvenile Instructor Office, 1881. [2nd ed., 1909; Rpt. 1966.]

> Hamblin narrated his autobiography to James Little. In 1870, Hamblin meets Powell and takes him to the Shivwits Indians to find out what happened to Dunn and the Howlands. He then accompanies Powell to Fort Defiance, crossing Lee's ferry and visiting the seven Hopi villages.

122. Hawkins, William W. [Letter]. *Adventures in the Canyons of the Colorado.* By William Wallace Bass. Grand Canyon, AZ: The Authors, 1920. 15–29.

> Hawkins was the cook on the 1869 expedition. His account was written a few months before he died in 1919. Hawkins portrays Powell as a callow, difficult man, and the reason Dunn and the Howlands left the expedition. Includes facsimile pages of the original manuscript letter. See also **#87** and **#123**.

123.——. "William Hawkins' Story." *Colorado River Controversies.* By Robert Brewster Stanton. New York: Dodd, Mead, 1932. 138–163.

> Hawkins wrote this account of his experience in 1907 in response to Stanton's request. He defends the reputation of the Howlands and Dunn against charges of cowardice and claims that he witnessed an attempt by an angry Powell to drown Dunn. See **#46**, **#87**, **#100**, **#122**, **#746**, and **#774**.

124. Hayden, Ferdinand Vandiveer. *Annual Reports of the U.S. Geological and Geographical Survey of the Territories.* Washington: GPO, 1867–1883.

> Twelve annual reports of the Hayden Survey, issued 1867–1878.

125. Hillers, John K., and Don D. Fowler. "*Photographed All the Best Scenery": Jack Hillers's* [sic] *Diary of the Powell Expeditions, 1871–1875.* U of Utah Publications in the American West, 9. Salt Lake City: U of Utah Pr, 1972.

> A lengthy introductory essay by Fowler provides background material, including Hillers' biography and an overview of his art. Fowler's transcription of Hillers' diary covers portions of 1871, 1872, 1873, and 1875 (see **#48**). Includes biographies and photographs of members of the 1871–1872 expedition and reproductions of a selection of Hillers' photographs.

126. Howland, Oramel G. "Letters of O. G. Howland to the *Rocky Mountain News*." Ed. William Culp Darrah. *UHQ* 15 (1947): 95–105.

> O. G., his brother Seneca, and William Dunn perished after leaving the river party at Separation Rapid on 28 Aug. 1869. The letters are dated 19 June, 30 June, and 1 Jul. See also **#746**.

127. Jones, Stephen Vandiver. "Journal of Stephen Vandiver Jones, April 21, 1871–December 14, 1872." Ed. Herbert Gregory. *UHQ* 16–17 (1948–1949): 19–174.

> Based on a typewritten copy in the library of the USGS, which in turn is a transcription of photocopies of the original journals at NYPL (see **#49**). Jones also wrote letters, which essentially duplicate his diary, to his hometown newspaper in Illinois, the *Lacon Home Journal* (12 Apr., 22 May, 10 June, 16 June, 20 July, and 31 Aug. 1871, and 10 July and 24 Aug. 1872).

128. Keplinger, Lewis Walter. "The First Ascent of Long's Peak." *The Trail* 8 (1915): 13–15.

> Inspired by Byers' article in *The Trail* (**#98**), Keplinger adds his own account of the ascent of Long's Peak by members of the 1868 expedition to the Rockies. See **#129**.

129.——. "The First Ascent of Long's Peak; Made by an Expedition Under Maj. J. W. Powell." *Collections of the Kansas State Historical Society (1915–1918)* 14 (1918): 340–353. [Rpt. in *The Trail* 12 (1919): 5–16, 27.]

> An expanded version of the 1915 article (**#128**). The scope is broadened to include more details regarding the entire 1868 expedition.

130. King, Clarence. "Annual Report of the Geological Exploration of the Fortieth Parallel." *Report of the Secretary of War, Report of the Chief Engineers*. Washington: GPO, 1871–1878.

> In addition to these annual reports, the King Survey published a series of final reports in seven monographs and one atlas, including King's classic *Systematic Geology* in 1878 (v. 1 in the series).

131. Lee, John Doyle. A Mormon Chronicle: The Diaries of John D. Lee, 1848–1876. Ed. Robert Glass Cleland and Juanita Brooks. Huntington Classics. San Marino, CA: Huntington Library, 2003.

> First published in 1955, this annotated transcription of Lee's five diaries from the Huntington Library collection chronicles life on the Mormon frontier. Lee was an early Mormon settler tried and executed

in 1877 for carrying out the murder of "gentiles" 20 years earlier at Mountain Meadows, Utah. Immediately following the massacre, Lee and his wives lived in exile at the junction of the Paria and Colorado Rivers. Here, as his diary records, they encountered Powell and members of his party in 1870 and again in 1872 and 1873.

132. Moran, Thomas. *Home-Thoughts From Afar: Letters of Thomas Moran to Mary Nimmo Moran*. Ed. Amy Bassford. Intro. by Fritiof Fryxell. East Hampton, NY: East Hampton Free Library, 1967.

A selection of letters written by Moran to his wife. Chapter 2, "1873: A Journey with Major John Wesley Powell" (pp. 25–42), contains several letters describing Moran's trek with Thompson, Hillers, and occasionally Powell, from Salt Lake City to the North Rim in 1873. Illustrated with Moran's art.

133. Powell, Walter Clement. "Journal of W. C. Powell, April 21, 1871–December 7, 1872." Ed. and transcr. Charles Kelly. *UHQ* 16–17 (1948–1949): 257–478.

Journal kept by Powell's cousin during the second Colorado River expedition.

134.——. "W. C. Powell's Account of the Hopi Towns." *UHQ* 16–17 (1948–1949): 479–490.

Reprint of an undated letter to the *Chicago Tribune* published 25 Feb. 1873.

135. Powell, Walter Henry. "Letter of W. H. Powell to the *Chicago Evening Journal*." Ed. William Culp Darrah. *UHQ* 15 (1947): 90–92.

The letter, written by Powell's youngest brother on 3 July 1869, describes the first two weeks of the voyage from Green River to Red Canyon.

136. Steward, John F. "Journal of John F. Steward, May 22–November 3, 1871." Ed. and transcr. William Culp Darrah. *UHQ* 16–17 (1948–1949): 180–251.

This publication of Steward's diary is from the copy at NYPL. See **#54**.

137. Sumner, John Colton. "Jack Sumner's Account." *Colorado River Controversies*. By Robert Brewster Stanton. New York: Dodd, Mead, 1932. 164–213.

Written in 1907, this account of the 1869 expedition disputes Powell's claim that Dunn and the Howland brothers left the party because they were afraid of a rapid. Stanton also reprints correspondence from Hawkins and Sumner regarding Powell's use of a life preserver. See **#137**, **#746**, and **#774**.

138.——. "J. C. Sumner's Journal, July 6–August 31, 1869." Ed. William Culp Darrah. *UHQ* 15 (1947): 113–124.

This publication of the second half of Sumner's diary is based on the copy in the Stanton collection at NYPL (**#8**). A copy of that copy, along with the field notes in Sumner's handwriting, are in the permanent collection of the Smithsonian (**#57**). The first half of the journal was published in *UHQ* 37 (**#140**). See also **#100** and **#746**.

139. Sumner, John Colton, and Don Lago. "Jack Sumner Looks Back." *boatman's quarterly review* 14.2 (2002): 40–41.

Lago introduces and transcribes a letter from Sumner to Lewis Keplinger. The letter in the Keplinger papers (**#50**) is consistent with Sumner's statements to Robert Stanton (**#137**).

140. Sumner, John Colton, and Otis R. Marston. "The Lost Journal of John Colton Sumner." *UHQ* 37 (1969): 173–189.

This first portion of Sumner's diary (24 May 1869–28 June 1869), here transcribed, edited and introduced by Marston, was mailed by Powell (see **#225**) from the Unita Indian agency on 6 July 1869 to the *St. Louis Missouri Democrat*, where it was published 28–29 Aug. 1869. The diary covers the journey from the Green River to the mouth of the Uintah. The second portion of the diary was published in *UHQ* 15 (**#138**).

141. Taylor, J. B. "In the Wesleyan, '63–'69." *Illinois Wesleyan Magazine* 5.1 (1900): 65–69.

In this reminiscence of his days at IWU, Taylor summarizes Powell's teaching thus: "Text books went to the wind."

142. Thompson, Almon Harris. "Diary of Almon Harris Thompson." Ed. and intro. Herbert E. Gregory. *UHQ* 7.1–7.3 (1939): 3–138.

All of Thompson's journals are published in this lengthy article, spanning the years 1871 to 1875 (see **#58**). Along with news articles about the Powell expedition published in the *Deseret Evening News* in 1871–1873, there is an excerpt from Powell's preliminary *Report of Explorations in 1873* (**#234**, **#235**).

143. Thompson, Ellen, and Beatrice Scheer Smith. "The 1872 Diary and Plant Collections of Ellen Powell Thompson." *UHQ* 62 (1994): 104–131.

Ellen's only diary from Utah covers mainly 5 Mar. to 16 May 1872. "Nell" Thompson collected plants on her travels through Utah and northern Arizona with husband A. H. Thompson. Her notes on the plants are sketchy, but the diary offers a wealth of details about daily life and her itinerary.

144. Watson, Elmo Scott. *The Professor Goes West; Illinois Wesleyan University—Reports of Major John Wesley Powell's Explorations, 1867–1874.* Bloomington: Illinois Wesleyan U Pr, 1954.

According to his foreword, Watson intended this publication to supplement a chapter in his earlier book, *The Illinois Wesleyan Story: 1850–1905* (**#783**). Drawing heavily on dispatches to newspapers, unpublished letters, and diaries written by expedition members, Watson documents Powell's 1867 and 1868 expeditions to the Rocky Mountains and the 1869 and 1872 Colorado River expeditions from the point of view of party members associated with IWU. Included are excerpts from fellow professor H. C. DeMotte's articles from the *Illinois Wesleyan Alumni Journal* (**#111–#116**) about his experience as a surveyor with the 1874 survey in Utah. See also **#101**.

145. Wheeler, George Montague. "Annual Report of the Geographical Surveys West of the One Hundredth Meridian." *Report of the Secretary of War, Report of the Chief of Engineers*. Washington: GPO, 1873–1879.

In addition to these reports issued for 1873–1884, the Wheeler Survey published a series of final reports, *Report Upon United States Geographical Surveys West of the One Hundredth Meridian*, 1875–1889.

146.——. *Report Upon United States Geographical Surveys West of the One Hundredth Meridian, Vol. I, Geographical Report*. Washington: GPO, 1889.

This volume is dedicated to the history of voyages and explorations of the western half of the United States. "Itinerary of Colorado Grand Cañon and River Trip of 1871" (pp. 147–171) presents a brief summary of explorations in the Colorado River region and mentions Powell's expeditions. "Memoir Upon the Voyages, Discoveries, Explorations, and Surveys to and at the West Coast of North America and Interior of the United States West of Mississippi River, from 1500 to 1880" (Appendix F) includes an overview of the work of the Powell Survey with bibliographical references to printed reports (pp. 713–718).

147. White, James. "Passage of the Great Canyon of the Colorado." *New Tracks in North America*. By William A. Bell. 2nd ed. London: Chapman and Hall, 1870. 435–453. [Rpt. Albuquerque, NM: Horn and Wallace, 1965.]

In this letter written in 1867, White claims to have taken an accidental and much-disputed ride on a raft through the Grand Canyon. Reprinted in Bass's *Adventures in the Canyons of the Colorado* (**#87**).

2. Bureau of American Ethnology

148. Bandelier, Adolph. *Pioneers in American Anthropology; the Bandelier-Morgan Letters 1873–1883*. Ed. and intro. Leslie A. White. Coronado Historical Series. Albuquerque: U of New Mexico Pr, 1940.

Lewis Morgan and his student Adolph Bandelier studied the social organization of peoples of ancient Mexico. In 1880, Bandelier began archaeological fieldwork in New Mexico for the Archaeological Institute of America. Most of these letters from Bandelier to Morgan are in Rush-Rhees Library. There are a number of references to Powell, whom Bandelier admired.

149. Morgan, Lewis Henry. *Ancient Society; or, Researches in the Line of Human Progress From Savagery Through Barbarism to Civilization*. Chicago: C. H. Kerr, 1877.

Powell was profoundly influenced by Morgan's theory of social evolution. In this two-volume work, Morgan identifies characteristic forms of government, family, and property in their progression from savagery to barbarism to civilization. Powell shared Morgan's belief that democracy would prevail by natural selection. Many republications and reprints.

150. Powers, Stephen, et al. *Stephen Powers, California's First Ethnologist; and, Letters of Stephen Powers to John Wesley Powell Concerning* Tribes of California. Contributions of the University of California Archaeological Research Facility 28. Berkeley: Archaeological Research Facility, Dept. of Anthropology, U of California, 1975.

Stephen Powers was a BAE ethnologist who wrote *Tribes of California*, v. 3 of the Contributions to American Ethnology series. Part 1 is "The Life of Stephen Powers," by Susan Park. Part 2, edited by R. F. Heizer, contains letters of Stephen Powers to John Wesley Powell concerning *Tribes of California*, and three letters from Powell to Powers written between 1880 and 1882. See **#34**.

3. U.S. Geological Society

151. King, Clarence. "First Annual Report [of the Director] of the United States Geological Survey." *First Annual Report of the United States Geological Survey, 1879–1880*. Washington: GPO, 1880.

> King, Director of the USGS for its first year, summarizes the work of the federal government in geology and the laws relevant to the establishment of the Geological Survey. He then reports on the first year of the survey's operation with reports from geologists in the field.

152. Ward, Lester Frank. *Dynamic Sociology, or Applied Social Science*. New York: Appleton, 1883.

> Ward dedicates this two-volume work to Powell, his mentor and close friend. Ward's thesis is that human progress is the result of a fundamental desire for happiness, which in turn is the stimulus for all social activities and systems. He also believed that if all citizens received a scientific education, then nature could be harnessed for the purpose of organizing a rational, progressive society in which wealth would be equally distributed. Powell reviewed Ward's book favorably in a lengthy essay (**#318**).

153.——. *Glimpses of the Cosmos*. New York and London: G. P. Putnam's Sons, 1913–1918.

> Ward's reminiscences of his own life and his observations on Powell and many of the prominent intellectuals of the late 19th century. See **#683** and **#878**.

4. Other

154. Agassiz, Alexander. *Letters and Recollections of Alexander Agassiz; With a Sketch of his Life and Work*. Ed. George R. Agassiz. Boston: Houghton Mifflin, 1913.

> This influential Harvard scientist became one of Powell's opponents.

155. Cushing, Frank Hamilton. *Cushing at Zuni: The Correspondence and Journals of Frank Hamilton Cushing, 1879–1884*. Albuquerque: U of New Mexico Pr, 1990.

> This BAE anthropologist employed unorthodox techniques in his study of the Zuni.

156. Garfield, James A. *The Diary of James A. Garfield.* Ed. Harry James Brown and Frederick D. Williams. East Lansing: Michigan State U Pr, 1967.

Four-volume transcription of the journal of one of Powell's strongest political supporters.

157. Garland, Hamlin. *Roadside Meetings.* New York: Macmillan, 1930.

In this collection of reminiscences, Garland encounters the ailing Powell in Washington in 1902 (pp. 361–363). In his honor, Garland pens a poem, "The Stricken Pioneer."

158. Grant, Ulysses S. *Memoirs and Selected Letters.* Ed. Mary Drake and William S. McFeeley. New York: Library of America, 1990.

Includes General Grant's recollections of the events at the Battle of Shiloh.

159. ——. *Papers of Ulysses S. Grant.* Ed. John Y. Simon. Carbondale and Edwardsville: Southern Illinois U Pr, 1967–.

This multivolume publication of Grant's papers is still in production. There are letters to and from Powell, as well as references to him and his expeditions, scattered throughout.

160. Ives, Joseph C. *Report Upon the Colorado River of the West, Explored in 1857 and 1858 by Lieutenant Joseph C. Ives.* 36th Cong., 1st Sess., H. Ex. Doc. 90. Washington: GPO, 1861. [Rpt. New York: Da Capo Pr, 1969.]

Report of an incomplete effort by Army engineer Ives to survey the Colorado River by steamship. John S. Newberry recorded the first significant observations of the geology. Survey artists Balduin Mollhausen and Baron von Egloffstein produced dramatic illustrations. For Powell's appreciation of Newberry's work, see *Exploration* (pp. 197–198).

161. Macomb, John N. *Report of the Exploring Expedition From Santa Fe, New Mexico, to the Junction of the Grand and Green Rivers of the Great Colorado of the West, in 1859, Under the Command of Capt. J. N. Macomb; With a Geological Report by Prof. J. S. Newberry.* Washington: GPO, 1876.

A report on a second, incomplete exploration of the Colorado River, this time by land. With Ives' report (**#160**), this was the best information available on the Colorado River prior to Powell's 1869 expedition.

162. Newcomb, Simon. *Reminiscences of an Astronomer.* Boston and Cambridge: Houghton, Mifflin and Riverside Pr, 1903.

Newcomb's autobiography provides an insider's view of the NAS as well as of political and social aspects of 19th-century science in Washington.

163. Stanton, Robert Brewster. "Through the Grand Cañon of the Colorado." *Scribner's Monthly* 8 (1890): 591–613.

An account of Stanton's failed first river expedition through the Grand Canyon in the spring of 1889 and the second successful survey in 1890. See also **#164** and **#724**.

164. Stanton, Robert Brewster, and Dwight L. Smith. *Down the Colorado.* Norman: U of Oklahoma Pr, 1965.

This account of Stanton's survey of the Colorado River in 1889–1890 is the first half of Stanton's massive manuscript, "The River and the Canyon," edited and introduced by Dwight Smith. It contains a detailed account of the Brown-Stanton river expeditions of 1889 and 1890, an overview of Colorado River expeditions and voyages, and Stanton's theory of the geologic history of the country. Stanton's second expedition was only the second complete exploration of the lower Colorado River. The second half was published earlier as *Colorado River Controversies* (**#774**).

165. Stewart, William Morris. *The Reminiscences of William M. Stewart of Nevada.* Ed. G. R. Brown. New York: Neale Pub Co, 1908.

An entertaining autobiography with no direct reference to Powell.

E. OTHER DOCUMENTS

1. Illinois Sources

166. University of Illinois. *Proceedings of the Board of Trustees.*

Then known as the Illinois Industrial Institution.

1867–1868: Powell's report "Rocky Mountain Expeditions" (**#218**); Cabinet grant to Powell for $500; and his election to professorship.

1868–1869: Powell's appointment to chair Natural History and his resignation.

1869–1870: Powell addresses the trustees.

167. Illinois Board of Education. *Proceedings.*

1867–1868: State appoints Powell as Curator of the Museum of Natural History Society; Powell gives Board a tour of the museum and an account of his labors in Colorado in 1867; authorization of $400 for purchases for 1868 expedition.

1871–1872: Resignation from his professorship at ISU.

168. Illinois General Assembly. *House Journal.*

1867: Powell's testimony before the General Assembly regarding appropriation for his first expedition.

2. Federal Sources

Bureau of American Ethnology Publications

For a comprehensive bibliography, see *The Bureau of American Ethnology* by Judd (#**814**).

169.BAE Annual Reports.

Each includes illustrated essays on basic researches in anthropology and a Director's Report. The first was published in 1881 for fiscal year 1879–1880.

170.BAE Bulletins.

Authorized in 1886 as a "cheaper form" than Contributions or Annual Reports. The original 24 were published until 1894, when authorization lapsed, then continued in 1900. They are primarily linguistic subjects—bibliographies, texts, and dictionaries.

171. Contributions to North American Ethnology.

The series began in 1877 as publications of the U.S. Geographical and Geological Survey of the Rocky Mountain Region (vols. 1–3). When the BAE was organized in 1881, vols. 6–10 were authorized, but the series discontinued in 1895 after vols. 1–7 and 9 had been completed.

172.BAE Introductions.

A series of pocket-sized leaflets, intended chiefly for the use of collaborators and correspondents of the BAE. Four were issued (1877–1880).

173.BAE Miscellaneous Publications.

Irregularly published, chiefly for internal use; they include proof sheets, maps, and reprints.

U.S. Geological Survey Publications

For a comprehensive bibliography of the first 20 years, see Warman's *Bibliography and index of the publications of the United States Geological Survey, 1880 to 1901* (**#979**). Reports, treatises, and maps of the King, Powell, Hayden, and Wheeler surveys are documented in Schmeckebier's *Catalogue* (**#976**).

174.USGS Annual Reports.

The annual reports of the USGS were published as separates but also form a part of the Secretary of the Interior's annual reports. Each volume contains a brief summary of the plans and operations of the survey and short administrative reports from chiefs of division, followed by scientific papers. Entries for director's reports are in Works of Powell.

175.USGS Bulletins.

In 1882, Powell initiated this series of shorter but more technical papers not represented in the annual report. Each is devoted to a specific topic and vary in length from five to several hundred pages.

176. Folios of the Geologic Atlas.

Inaugurated in 1894 but based on earlier work of the Survey under Powell's direction. Each folio describes the topography, geology, and mineral resources of a specific area.

177.USGS Monographs.

A series of book-length reports on scientific investigations, sometimes accompanied by atlases. Powell added paleontology to King's original scope of economic and general geology.

178.USGS Reports on Mineral Resources.

Powell began this series of compiled statistics related to mines and mining in 1882.

Other Federal Publications

179. *Congressional Globe.*

The *Globe* is an important source for tracing debates over the purposes, nature, and impact of 19th-century exploration of the western United States. See years 1868–1873, especially May 1868 regarding authorization for issuing government supplies to Powell and July 1870 for appropriations for the second Colorado River expedition.

180. *Congressional Record.*

See years 1873–1903, especially the following:

1879: NAS report on public lands.
1886: Rep. Hilary Herbert's charges against Powell.
1888: First extended discussion of the reclamation of arid lands in Congress.
1890: Reprint of Powell's interview with the *Washington Star* regarding irrigation and the ruling prohibiting irrigable lands for entry/claims.
1892: Debate over Powell's budgets; attacks by Stewart and other Westerners.

181. Smithsonian Institution Annual Reports.

These contain reports on the Powell Survey and the BAE under Powell, 1867–1903. Reports by Powell are referenced in Chapter 2.

182. U.S. Bureau of Indian Affairs. Commissioner of Indian Affairs Annual Reports.

Years 1873–1875 contain information on Powell's activities in the Southwest. Reports by Powell are referenced in Chapter 2.

183. U. S. Dept. of Interior Annual Reports.

These contain annual reports of the USGS and the earlier Powell Survey beginning 1867.

F. CONTEMPORARY NEWSPAPERS AND PERIODICALS

1. Newspapers

184.*Daily Pantagraph.* Bloomington, IL.

185.*The Weekly Leader.* Bloomington IL.

186.*Rocky Mountain News.* Denver, CO.

187.*Chicago Tribune.* Chicago, IL.

188.*Deseret Evening News.* Salt Lake City, UT.

189.*The New York Times.* New York.

190.*Home Journal.* Lacon, IL.

191.*New York Herald.* New York.

Powell's activities associated with IWU, ISU, and the Illinois Natural History Society were reported in the Bloomington (Illinois) *Daily Pantagraph*. The Rocky Mountain expeditions of 1867 and 1868 received considerable local coverage in both the *Pantagraph* and the Denver *Rocky Mountain News*. Several members of the expeditions wrote letters to the *Pantagraph*, which also gave extensive coverage to the 1869 Colorado River exploration. Reprints in **#144**.

The Weekly Leader, another Bloomington paper, published an interview with *Emma* Powell in July 1869 regarding the reported loss of the Powell party. Transcripts and copies of articles from the Bloomington papers are held by the McLean County Historical Society in Bloomington, Illinois.

The 1869 and 1871–1872, Colorado River expeditions received national attention, with notable coverage in the *Chicago Tribune,* the Salt Lake City *Deseret Evening News,* the Denver *Rocky Mountain News*, and *The New York Times*. In addition, members of the expeditions contributed letters to their local newspapers, such as S. V. Jones's letters to the Lacon (Illinois) *Home Journal*. See entries **#92**, **#126**, **#127**, **#135**, and **#220–#225** for individual letters. Reprints in **#102**, **#103**, **#213**, and **#746**.

The *New York Herald* carried sensational coverage of the "Cope-Marsh Controversy" in January 1890 (see **#405**). See also Washington newspapers for political coverage of Powell's run-ins with Congress as well as reports of activities of the many clubs and organizations to which Powell belonged.

2. Journals and Magazines

Publications of the Anthropological Society of Washington

Abstracts, remarks, and texts of papers by Powell as recorded in these publications are cited in Chapter 2.

192. *Transactions of the Anthropological Society of Washington.*

First published in 1882, v. 1 (1879–1882) contains a lengthy synopsis of the first three years of meetings, with it's the ASW's constitutions and Powell's presidential address of the year. Publication continued with two more volumes through 1885. The Smithsonian Institution supported publication of *Transactions*, which also published v. 1 in *Smithsonian Miscellaneous Collections*, v. 25 (**#195**). Reprinted by Kraus in 1971.

193. *Abstract of Transactions of the Anthropological Society of Washington, D.C.; With the Annual Address of the President, for the First Year, Ending January 20, 1880, and for the Second Year, Ending January 18, 1881.*

This publication was prepared by Powell and printed in 1881 at the joint expense of Powell and the Society. It was reprinted by the Smithsonian Institution in its *Miscellaneous Collections*, v. 25 (**#195**). See also **#288**.

194. *American Anthropologist.*

Continues *Transactions*, but articles are full-length rather than abstracts and resumes of papers presented at meetings of the ASW. Publication began with v. 1 in 1888.

195. *Smithsonian Miscellaneous Collections*

The Smithsonian Institution reprinted several publications of Washington's scientific organizations in this series. In subsequent annotations, these publications are abbreviated as *SMC*, followed by volume number.

Vol. 20 (1881): *Bulletin of the PSW*, v. 1–3.

Vol. 25 (1883): *Bulletin of the PSW*, v. 4–5; *Transactions of the ASW*, v. 1; *Abstract of Transactions of the ASW*; *Proceedings of the BSW*, v. 1.

Vol. 33 (1888): *Bulletin of the PSW*, v. 6–10, with index to all volumes; *Memorial Proceedings in Honor of Professor Baird.*

Other Journals

196.*American Geologist.*

197.*American Journal of Science.*

198.*Appleton's Monthly.*

199.*Bulletin of the Philosophical Society of Washington.*

200.*Century Magazine.*

201.*Garden and Forest.*

202.*Harper's Monthly.*

203.*Illinois Wesleyan University Alumni Journal.*

204.*Irrigation Age.*

205.*Journal of the American Geographical Society.*

206.*Monist.*

207.*Nation.*

208.*North American Review.*

209.*Open Court.*

210.*Overland Monthly.*

211.*Popular Science Monthly.*

212.*Science.*

Chapter 2

Published Writings and Addresses of John Wesley Powell

A. COLLECTIONS

213. Powell, John Wesley, and William Eno deBuys. *Seeing Things Whole: The Essential John Wesley Powell.* Washington: Island Press/Shearwater Books, 2001.

> DeBuys compiles, edits, and annotates an essentially chronological selection of Powell's published works. Although the choices cover the range from letters to the 1869 *Chicago Tribune* to articles on ethnology, and though each of the three articles on irrigation from *Century Magazine* is reprinted, the emphasis here is on *Report on the Lands* and related works. DeBuys provides expert commentary in introductory essays and notes. The chapter "Works by John Wesley Powell" is a chronological bibliography of major works (pp. 359–367).

214. Powell, John Wesley, and George Crossette. *Selected Prose of John Wesley Powell.* Boston: David R. Godine, 1970.

> This compilation of selections from Powell's lesser-known nontechnical writings, edited and introduced by Crossette, represents the broad range of Powell's knowledge and interests. Most selections are short excerpts from journal articles and addresses. The selection is organized by subject (Language & Music, Myth & Folklore, etc.) and includes a section on "Aphorisms & Analogies."

B. INDIVIDUAL WORKS

Within each year, citations appear by type of publication in the following order: Monographs (books, whole documents, pamphlets and offprints); Articles (periodical articles and essays, chapters, and transcripts of addresses in whole works by others); Abstracts and Remarks; and Reprints (used only for works published significantly later than the original).

1867

Monographs, Pamphlets, Whole Documents

215.*Scientific Expedition to the Rocky Mountains. Preliminary Report of Prof. J. W. Powell to the Illinois State Board of Education*. Peoria, IL: N. C. Nason, Printer, 1867.

> Pamphlet edition (4 pp.) of Powell's report, presented at the board's Regular Meeting, 17–18 Dec. 1867 (see also **#217**). The report covers the first of two expeditions to Colorado with parties of amateur scientists and students. Powell documents President Grant's assistance in the form of military escort and supplies from Fort Laramie and acknowledges financial support from the Illinois Board of Education, the Industrial University (University of Illinois), and the Chicago Academy of Science. He describes the ascent of Pike's Peak and states his intent to return the next year to the headwaters of the Colorado River.

Articles, Selections

216."Museum. Professor Powell's Statement." *Reports Made to the General Assembly of Illinois at Its Twenty-Fifth Session*. Vol. 2. Springfield, IL: Baker & Bailhace; 1867. 233–235 [241–243].

> Powell presents a detailed plan for the museum of the Illinois State Natural History Society in Normal and a request to the Illinois Board of Education for financial assistance. Powell argues the society needs a curator to take charge of the museum, superintend researches and collections, and distribute collections to schools and universities throughout the state. Further, "All civilized nations deem it wise to foster such institutions." This statement is part of the report of the president of the Normal University (ISU) in the "Sixth Biennial Report of the Superintendent of Public Instruction of the State of Illinois for 1865–1866."

217."Scientific Expedition to the Rocky Mountains. Preliminary Report of Prof. J. W. Powell to the Illinois State Board of Education." *Proceedings of the Board of Education of the State of Illinois.* Peoria, IL: Printed by N. C. Nason, Office of the Illinois Teacher, 1867. 9–11.

See **#215** for annotation.

1868

Abstracts, Remarks

218."Major Powell's Report." *Proceedings of the Board of Trustees [University of Illinois] for 1867–1868.* N.p.: n.p., 1868. 120.

Summary of Powell's report on the Rocky Mountain Expedition at the second annual meeting on 11 Mar. 1868. The collection is being classified and prepared for presentation. He announces plans for another expedition.

1869

Articles

In addition to journals, members of both Colorado River parties wrote letters to newspapers. Powell's letters, listed in chronological order below, form an especially important body of primary documents from the 1869 expedition. Letters published in the *Chicago Tribune* are reprinted in *UHQ* 15 (**#568**), *Professor Goes West* (**#144**), and *First through Grand Canyon* (**#746**).

219."The Exploring Expedition." Bloomington *Daily Pantagraph*, 12 Feb. 1869.

Publication of a letter written to Dr. Richard Edwards, president of ISU, from winter camp on the White River, on 26 Jan. 1869. Powell refers to exploring the canyon of the Green River, which was said to be impassable, and looks forward to "making the passage of the 'Grand Cañon' of the Colorado."

220."The Powell Expedition." *Chicago Tribune*, 29 May 1869.

Letter dated 24 May 1869 from Green River, Wyoming. Powell gives a full account of how the expedition was financed. He states his goal is to study the geography and geology of the Colorado River valley and the unexplored "Grand Cañon."

221. Letters. *Chicago Tribune*, 19 July 1869.

Two letters dated 2–6 June and 7 June 1869.The first letter traces events from 24 May at Flaming Gorge through the passage of Red Canyon 2. The second covers 4 June at Brown's Hole to 6 June at the mouth of Lodore Canyon.

222. Letter [to Dr. Richard Edwards]. Bloomington *Daily Pantagraph*, 19 July 1869.

Written from the mouth of the Uinta River, where the party camped for several days, and dated 29 June 1869. Powell reports losing his papers in the wreck of one of his boats, the No-Name. All party members are in good health and good spirits. The letter was published in other newspapers, including the *Chicago Tribune* (20 July 1869) and *The New York Times* (23 July 1869). Reprinted in *Seeing Things Whole* (**#213**).

223. Letter [to Dr. Henry Wing]. St. Louis *Missouri Republican*, 22 July 1869.

Dated 29 June 1869, from the Uinta. Similar to preceding letter to Edwards (**#222**).

224. Letters. *Chicago Tribune*, 20 Aug. 1869.

Two letters dated 18–20 June and 23 June 1869, from the entrance of Lodore Canyon through Echo Park, with news of the wreck of the No-Name. Reprinted in *Seeing Things Whole* (**#213**).

225."From Col. Powell." St. Louis *Missouri Democrat* 24 Aug. 1869.

Letter dated 6 July 1869 transmitting John Sumner's journal. This letter prefaced the publication of Sumner's journal (**#140**) in this and the next issue.

1870

Articles

226."Letter to James G. Blaine." *Exploration of* [sic] *Colorado River: Letter From the Secretary of the Interior Transmitting an Estimate of Appropriations to Prosecute to Completion the Geological and Topographical Exploration of the Colorado River and Its Tributaries.* 41st Cong., 2nd Sess., H. Ex. Doc. 280 Serial 1426. Washington: GPO, 1870. 1–2.

> Powell requests $12,000 to complete the geographical exploration of the Colorado River between the junction of the Grand and Green Rivers and the Virgin River. The money is to be used for personnel, instruments, and a pack train. His request was granted, essentially creating what would officially become the U.S. Geographical and Geological Survey of the Rocky Mountain Region, also known as the Powell Survey. See Petition (**#228**).

227."Major J. W. Powell's Report on His Explorations of the Rio Colorado in 1869." *New Tracks in North America.* 2nd ed. Ed. William A. Bell. London: Chapman and Hall; New York: Scribner, Welford, 1870. Appendix D, 559–564. [Rpt. New Mexico: Horn and Wallace, 1965.]

> First complete published report of Powell's 1869 expedition (though Bell notes that brief accounts had appeared in "some of the American papers"). Powell observes the geological processes that subsequently formed his theory of topographic development at work in the Grand Canyon: uplift, erosion, and sedimentation. *New Tracks* includes James White's "Passage of the Great Canon of the Colorado." Reprinted in **#569** and **#576**.

228. Petition to the Senate and House of Representatives. *Exploration of the Colorado River; Letter From the Secretary of War Transmitting a Report of the Chief of Engineers Upon the Proposed Continuation of the Exploration of the Colorado River.* 41st Cong., 2nd Sess., H. Ex. Doc. 281 Serial 1426.Washington: GPO, 1870. 2.

> Same request for funding as above (**#226**), this time endorsed by Secretary Belknap.

1871

Articles

229. Letter. *The National Elgin Watch Company Illustrated Almanac 1871.* Chicago: National Watch Co., [1871?]. 19. [Rpt. Northbrook, IL: Bernard J. Edwards, 1977.]

> The letter, in an advertisement for Elgin watches, is dated 20 June 1870 and addressed to T. M. Avery, President of the National Watch Company. The return address is Illinois State Natural History Society in Bloomington. The endorsement gives personal testimonial for the Elgin watch Powell used during the 1869 expedition. See also facsimile publication (**#582**).

1872

Monographs

230.*Survey of the Colorado River of the West; Letter From the Secretary of the Smithsonian Institution, Transmitting Report Preliminary for Continuing the Survey of the Colorado of the West and Its Tributaries.* 42nd Cong., 2nd Sess., H. Misc. Doc. 173 Serial 1526. Washington: GPO, 1872.

> In this first preliminary report of the Powell Survey, Powell describes in some detail the expedition's movements and activities from August 1870 through the end of 1871. With funds remaining from Congressional appropriations, he will complete the exploration of the Grand Canyon. He requests continued funding to expand the scope of his survey to include the plateau and canyon country of Utah, Arizona, and parts of western Colorado and New Mexico.

1873

Monographs

231.*Report of the Survey of the Colorado of the West. Letter From the Secretary of the Smithsonian Institution, Transmitting a Report of the Survey of the Colorado of the West, and Its Tributaries.* 42nd Cong., 3rd Sess., H. Misc. Doc. 76 Serial 1572. Washington: GPO, 1873.

> This second report presents a summary of the work completed by the Powell Survey in 1872 and an estimate of the cost of continuing the

survey. Powell classifies the valleys of the region in two orders, longitudinal and transverse, and each of these in three "varieties." A part of this report is devoted to ethnographic studies.

Articles

232."Some Remarks on the Geological Structure of a District of Country Lying to the North of the Grand Cañon of the Colorado." *American Journal of Science and Arts* 3rd ser. 5 (1873): 456–465.

Powell's first scientific publication, in which he systematically describes the "great geographical stairway" north of the Grand Canyon. The most remarkable features are the deep, narrow canyons and the long lines of cliffs. Powell adopts local Indian names or English equivalents for topographical features and proposes a classification scheme for the valleys. See **#241**.

Warman separate not seen.

1874

Monographs

233.*Indians West of the Rocky Mountains: Statement of Major J. W. Powell Made Before the Committee on Indian Affairs as to the Condition of the Indian Tribes West of the Rocky Mountains.* 43rd Cong., 1st Sess., H. Misc. Doc. 86 Serial 1618. Washington: GPO, 1874.

This pamphlet contains Powell's testimony regarding the report of Special Commissioners Powell and George W. Ingalls (see **#236**). His testimony contains the substance of the report, with elaboration on some points.

234.*Professor Powell's Report on the Survey of the Colorado of the West. Letter From the Secretary of the Smithsonian Institution, Transmitting a Report of Professor Powell on the Survey of the Colorado River of the West and Its Tributaries, &c., &c.* 43rd Cong., 1st Sess., H. Misc. Doc 265 Serial 1621. Washington: GPO, 1874.

Summary of the work of the Powell Survey to date, including the 1869 expedition. Powell details findings in the areas of topography and structural geology and addresses paleontology and economic geology. A lengthy section on ethnography contains an account of the mythology of the Numic Indians and gives several examples of the large collection of poems documented by Powell. See **#235**.

235. *Report of Explorations in 1873 of the Colorado of the West and Its Tributaries*. Washington: GPO, 1874.

Pamphlet edition of *Professor Powell's Report on the Survey of the Colorado of the West* (**#234**). Printed at the head of the text: "Report of Professor Powell on the Survey of the Colorado of the West."

236. *Report of Special Commissioners J. W. Powell and G. W. Ingalls on the Condition of the Ute Indians of Utah; the Paiutes of Utah, Northern Arizona, Southern Nevada, and Southeastern California; the Go-Si-Utes of Utah and Nevada; the Northwestern Shoshones of Idaho and Utah; and the Western Shoshones of Nevada; and Report Concerning Claims of Settlers in the Mo-a-Pa Valley, Southeastern Nevada.* Washington: GPO, 1874.

Powell and co-commissioner George W. Ingalls were appointed to make a full study, with recommendations, of the native peoples of the Colorado Plateau and Great Basin in 1873. Their report is a summary of conditions of the Indians, a census of the "tribes," and recommendations for their removal to selected reservation sites. The commissioners propose dividing the peoples into four groups and concentrating them on four large reservations, noting that "their hunting grounds have been spoiled, their favorite valleys are occupied by white men." The report recommends that Indians give up claims on a half million square miles and government handouts in exchange for legal title to 10,000 square miles, schools, the goods and knowledge to farm and create their own clothing and tools, and medicine. The presence of troops should be removed. Additional material includes an estimate of appropriations for support of the "Indians heretofore described" for fiscal years 1873–1875 and two maps. Appended with a list of claims made by squatters who were displaced by creation of the Muddy River Reservation. This is a pamphlet edition (36 pp.) of the report in *Annual Report of the Commissioner of Indian Affairs for 1873* (**#239**). See also **#237**.

237. *Ute, Pai-Ute, Go-Si Ute and Shoshone Indians: Letter From the Acting Secretary of the Interior, in Relation to the Condition and Wants of* the *Ute Indians of Utah; the Pai-Utes of Utah, Northern Arizona, Southern Nevada, and Southeastern California; the Go-Si Utes of Utah and Nevada; the Northwestern Shoshones of Idaho and Utah, and the Western Shoshones of Nevada*. 43rd Cong. 1st Sess., H. Ex. Doc. 157 Serial 1610. Washington: GPO, 1874.

Same as **#236**, but without the claims of settlers. Reprinted in *Anthropology of the Numa* (**#29**).

Articles

238."The Colorado Cañons." *Tribune Popular Science*. Boston: Henry L. Shepard, 1874. Pt. III, 9–10.

The volume is a compilation of articles from the *New York Tribune* series of "Extras." This article, originally published in the *New York Tribune, Extra No. 19, Lectures and Letters*, is an excerpt from a paper presented in 1874 at the NAS's annual meeting. The text is from *Explorations*.

239."Report of J. W. Powell and G. W. Ingalls." *Annual Report of the Commissioner of Indian Affairs to the Secretary of the Interior for the Year 1873*. Washington: GPO, 1874. 41–74.

This report is labeled "B" in the papers accompanying the commissioner's report. See **#236** and **#237**.

240. Statements. *Geographical and Geological Surveys West of the Mississippi*. 43rd Cong., 1st Sess., H. Rept. 612 Serial 1626. Washington: GPO, 1874. 9–10, 46–56.

In response to Congressional investigations into the extent of overlap among the four existing surveys, the House Committee on Public Lands (known as the Townsend Committee) held hearings in 1874. This publication of testimony and findings contains letters and statements from Powell speaking on behalf of his own survey. On 24 Apr. 1874, Powell submits a statement in which he calls for consolidating all surveys, including military surveys, under one civilian department (9–10). On 18 May, Powell presents a list of personnel and finances for the Colorado River Exploring Expedition, summarizes its work in topography, and, for the first time, states his opinion that the total water supply of the West would always be too small to irrigate all the land (46–56). The committee recommended no action be taken.

Abstracts, Remarks

241."Remarks on the Structural Geology of the Valley of the Colorado of the West." *Bulletin of the Philosophical Society of Washington, 1871–1874* 1 (1874): 48–51.

Abstract of an address delivered 9 Mar. 1872. Primarily concerned with the drainage system, Powell proposes a classification of the types of valleys within the region. Based on the text of Powell's 1873 article

"Some Remarks on the Geological Structure of a District of Country Lying to the North of the Grand Canon of the Colorado" (**#232**).

1875

Monographs

First printed as a Congressional document (43rd Cong., 1st Sess., H. Misc. Doc. 300 Serial 1622), *Exploration of the Colorado River of the West and Its Tributaries* (referred to as *Exploration* in annotations) was written at the urging of then-senator James Garfield. Portions appeared nearly simultaneously in *Scribner's Monthly* (**#243**, **#244**, **#246**) and *Popular Science* (**#247**) in 1875. Revised and augmented with additional text and copious illustrations, it was republished for a popular audience as *Canyons of the Colorado* in 1895 (**#466**). In one form or another, *Exploration* has been almost continuously reprinted and remains one of the most popular documents of 19th-century exploration of the American West. It reads well as an adventure story, nature writing, popular science, or autobiography. Powell failed to mention the men of the 1871–1872 Colorado River journey, though he wove incidents from both expeditions into this single narrative of the 1869 trip from the Green River to the Virgin River, thus earning the ire of several members of the second trip. To this day, the work ignites controversy and contention among historians and Colorado River rafters, in particular because of Powell's rendition of the events at Separation Rapid, where William Dunn and brothers Oramel and Seneca Howland left the river party on 28 Aug. 1869. They were later found murdered.

242.*Exploration of the Colorado River of the West and Its Tributaries Explored in 1869, 1870, 1871, and 1872 Under the Direction of the Secretary of the Smithsonian Institution.* Washington: GPO, 1875.

> In the first eight chapters of Part 1, "The History of the Exploration of the Canyons of the Colorado," Powell combines the 1869 and 1871–1872 Colorado River journeys into one narrative. Chapter 9 of Part 1 recounts the 1870 journey through Zion and the Arizona Strip to Powell's meeting with Shivwits. The section "Report on a Trip to the Mouth of the Dirty Devil River," by A. H. Thompson, recounts incidents from May–July 1872, taken from his diary (see **#58**). Part 2 is a scientific treatise by Powell, "On the Physical Features of the Valley of the Colorado" (**#247**), which opened new fields for geological investigation. The last article and two addenda in Part 3 were tacked onto this report when printed and have no relationship to it. There are 80 illustrations, many by Thomas Moran. Includes two maps in a pocket: "Green River from the Union Pacific Rail Road to the Mouth of White River (1873)" and "Profile of the Green River and Colorado River of the West...."

Much of the text from Powell's letters to the *Chicago Tribune* was used verbatim in this report. (see **#220**, **#221**, **#224**). Reprinted in **#566**, **#570**, **#571**, **#573**, **#587**, **#589**, and **#591**.

Articles

243."The Ancient Province of Tusayan." *Scribner's Monthly* 11 (1875): 193–213.

This is the concluding article of the *Scribner's* series (see **#244**, **#246**). It is an account of Powell's journey in the fall of 1870 through the Kanab canyons and Mormon settlements, then across the Kaibab Plateau with Jacob Hamblin to the Indian town of Oraibi. Includes detailed descriptions of Navajo life and Hopi architecture, customs, and legends. Powell cites linguistic evidence to explain the origin of the Hopi. Illustrated with woodcuts by Moran and a map. Reprinted as *Hopi Villages* (**#580**) and in *Seeing Things Whole* (**#213**). See also **#584**.

244."The Cañons of the Colorado." *Scribner's Monthly* 9 (1875): 293–310, 394–409, 523–537.

The first three parts of this series (with **#243**, **#246**), is essentially the same as Powell's report in Part 1 of *Exploration*, 24 May–20 Sept. 1869. He tells the story behind the 1869 expedition, with details about personnel, and presents a chronological account of the journey from the Green River to the Virgin River. Numerous woodcuts by Moran were commissioned for the series. See also **#584**.

245."Classification of Indian Tribes in Alaska." *Annual Report of the Commissioner of Indian Affairs for the Year 1875.* Washington: GPO, 1875. 202.

At the request of Edward Smith, Powell provides information on his efforts to collect linguistic data for the purpose of classifying Indian tribes. Powell reports the work of Stephen Powers in California and transmits a report by William Dall, "Abstract of Population of the Native Tribes of Alaska."

246. "An Overland Trip to the Grand Cañon." *Scribner's Monthly* 10 (1875): 659–678.

Fourth part of the series (with **#243**, **#244**). Most of this article appears in slightly different form in *Exploration,* in which Powell describes the journey from Salt Lake City to the Grand Canyon in 1870. Regarding Indians, he is sympathetic to their complex social problems, but thinks they should be gradually absorbed into "more civilized communities." Includes a passage in which the chief of the Shivwits acknowledges killing three white men (presumably Dunn and the Howland brothers) and Jacob Hamblin is introduced. Illustrated with a map of the Grand Canyon and woodcuts by Moran. Reprinted in *Seeing Things Whole* (**#213**) and in **#581** and **#584**

247."Physical Features of the Colorado Valley." *Popular Science Monthly* 7 (1875): 385–399, 531–542, 670–680.

Excerpted from Part 2 of *Exploration*. With details about valleys, badlands, the Uinta Mountains, water sculpture, and records of ancient lands, Powell attempts to convey the unique features of the region and introduce complex geological concepts to a general audience. The term "base level of erosion" is introduced.

248."Survey Under Professor Powell." *Annual Report of the Board of Regents of the Smithsonian Institution ... 1874*. Washington: GPO, 1875. 40–42.

The report is unsigned. Warman credits this to Powell, but his authorship cannot be verified. The report covers the work of the survey for 1874. Topographical work in the Colorado River basin was organized into two parties, one under A. H. Thompson and the other under Powell. Ethnological work included the study of Gosiute religious ceremonies.

1876

Monographs

249.*Report on the Geology of the Eastern Portion of the Uintah Mountains and a Region of Country Adjacent Thereto*. Washington: GPO, 1876.

Based on his fieldwork in 1874–1875, Powell interprets the geological history of the Colorado River region. This work and Part 2 of *Exploration* are the main texts for Powell's theory of how erosion and upheaval affect topography. He explains his concepts of "superimposed" and "antecedent" drainage and "base level of erosion." Published with companion atlas (**#250**).

250. *Atlas Accompanying the Report on the Geology of the Eastern Portion of the Uintah Mountains and a Region of Country Adjacent Thereto.* New York: Julius Bien, Lith., 1876.

> Geological and surface feature maps of the Green River dated 1873. The plates are sections and diagrams of the Uinta Plateau, the Uinta Mountains, the Yampa Plateau, the Aspen Mountains, and the Park Mountains. See **#249**.

Articles

251. "Types of Orographic Structures." *American Journal of Science and Arts* 3rd ser. 12 (1876): 414–428.

> A description of the 11 types of structures in the country north of the Grand Canyon, as categorized by Powell. Extracted from *Report on the Geology of the Eastern Portion of the Uintah Mountains ...* (**#249**).

1877

Monographs

252. *Alphabet to Accompany "Introduction to the Study of Indian Languages."* Washington: GPO, 1877.

> This four-page pamphlet was issued by the BAE to accompany the first edition of *Introduction to the Study of Indian Languages* (**#253**).

253. *Introduction to the Study of Indian Languages, With Words, Phrases, and Sentences to Be Collected.* Washington: GPO, 1877.

> Based on the earlier work of George Gibbs for the Smithsonian Institution, this manual is essentially a code of instructions for BAE fieldworkers who collected linguistic material. Common words and phrases are listed in English; then, equivalent Indian words are transliterated into the Roman alphabet and placed in parallel lists so languages can be compared and grouped into families. The introduction (pp. 3–7) consists primarily of a paper by W. D. Whitney, "On the Alphabet," and is followed by 96 pages of charts arranged in 24 schedules, with blank pages for collecting data. This was intended to be a chapter in a larger work, a "Manual of North American Ethnology," which was never completed. Revised and much expanded in 1880 (**#276**). Distributed with *Alphabet* (**#252**). See also **#590**.

254.*Outlines of the Philosophy of the North American Indians; Read Before the American Geographical Society, at Chickering Hall, December 29, 1876.* New York: Douglas Taylor, 1877.

In this address, Powell for the first time describes his approach to the philosophy and belief systems of Indians. Here Powell is clearly influenced by the anthropologist Lewis Henry Morgan, whose *Ancient Society* (**#149**) identified three evolutionary stages of human history: savage, barbaric, and civilized. Powell assumes Morgan's approach to anthropology and philosophy. Accordingly, human societies evolve in stages comparable to the life of an individual: "savagery is ethnic childhood" and "civilization" is adulthood. Powell describes the "cosmology, theology, religion and mythology" of the "native races." To the "savage" mind, everything is known or has an explanation. Republished as "A discourse on the philosophy of the North American Indians" (**#263**).

255.*Report on the Geographical and Geological Survey of the Rocky Mountain Region.* Washington: GPO, 1877.

Powell communicates the fieldwork of the Powell Survey for 1876 and 1877 and the office work of fiscal year 1876–1877. The first part conveys the topographic and geological work, including a classification of lands by G. K. Gilbert. The second part comments on irrigable lands. Here Powell is working on computations to determine how much of the arid West is redeemable by irrigation. He mentions the importance of fire prevention and a survey of mineral lands. Accompanying map is "Index Map for the Physical Atlas of the Western Portion of the United States in Process of Construction by the Interior Department." This is a pamphlet edition of Powell's 1877 report, published in annual reports of the Commissioner of Indian Affairs (**#258**) and the Smithsonian (**#264**). Published as an article in 1878 (**#265**).

Articles

256."Appendix: Linguistics." *Tribes of California.* By Stephen Powers. Contributions to North American Ethnology 3. Washington: GPO, 1877. 439–613.

Comparative vocabularies for each of the language families of the California Indians, in the form of tables.

257."Preface." *Tribes of the Extreme Northwest.* By W. H. Dall. Contributions to North American Ethnology 1. Washington: GPO, 1877. v–ix.

> Powell relays the history behind this first publication in the series, including reprints of correspondence between Powell and Joseph Henry, in which Henry transfers material on Indian linguistics to Powell for publication by the Powell Survey. This publication also includes *Tribes of Western Washington and Northwestern Oregon*, by George Gibbs.

258."Report on the Geographical and Geological Survey of the Rocky Mountain Region." *Report of the Commissioner of Indian Affairs for 1876.* 45th Cong., 2nd Sess., H. Ex. Doc. 1, Pt. 5 Serial 1800. Washington: GPO, 1877. 789–805.

> See **#255** for annotation. See also **#264** and **#265**.

1878

Monographs

259.*Geological and Geographical Surveys. Letter From the Secretary of the Interior, Transmitting a Report of Professor Powell in Regard to Surveys.* 45th Cong., 2nd Sess., H. Ex. Doc. 80 Serial 1809. Washington: GPO, 1878.

> In the spring of 1878, the House Appropriations Committee directed Secretary of Interior Schurz to submit information regarding the possible consolidation of the four geological surveys. Powell, Hayden, and Wheeler each submitted reports on their surveys. In this report, Powell details the work of his surveys, defines areas of duplication among all the surveys, and outlines the achievement of each. There is also a list of publications of the Powell surveys and "Chart Exhibiting the Areas Surveyed by King, Wheeler, Hayden, Jenney, and Powell." Ultimately, the Appropriations Committee proposed that the NAS be authorized to review the whole subject of unification of the surveys and submit a report with recommendations to Congress (see **#725**).

260.*Report on the Lands of the Arid Region of the United States, With a More Detailed Account of the Lands of Utah.* Washington: GPO, 1878.

> One of the reports of the Powell Survey and a seminal work. In the preface and the first two chapters, Powell lays out a plan for the settlement of the entire West. Eastern institutions for the settlement, division, and governance of land will not work in the arid West. Instead, Powell proposes that public lands be surveyed and classified

according to their economic potential. Where rainfall is inadequate for agriculture, increase the homesteads from 160 to 2,560 acres or more to support cattle or sheep herds. Where irrigation is possible, reduce homesteads to 80 acres or less and provide federal aid to build dams that are too costly for individual settlers to finance. Land use and distribution should be based on the properties of the land itself, especially in regard to water resources.

The report is accompanied by a bill to authorize organization of irrigation districts for homestead settlement of public lands and a bill to organize pasturage districts. Chapters with supporting content are contributed by G. K. Gilbert, C. E. Dutton, A. H. Thompson, and Willis Drummond, Jr. Accompanying maps are "Rain Chart of the United States"; "Map of the Utah Territory"; and "Map of the United States exhibiting the grants of lands made by the general government to aid in the construction of railroads an wagon loads 1878." Also published as a Congressional document (45th Cong., 2nd Sess., H. Ex. Doc. 73). A second edition was published in 1879 (**#270**).

261.*Report on the Methods of Surveying the Public Domain.* Washington: GPO, 1878.

At the request of the NAS, Powell forwarded this formal report in which he presents his methods and a detailed account of expenses of his surveys. He recommends a survey to classify the public lands and calls for the consolidation of the various Western surveys under one civilian authority. Powell states, "The present multiplication of organizations is unscientific, excessively expensive, and altogether vicious, preventing comprehensive, thorough and honest research, stimulating unhealthy rivalry, and leading to the production of sensational and briefly popular rather than solid and enduring results." The NAS incorporated Powell's program virtually intact in its report to the Secretary of the Interior.

This is a pamphlet edition of the same in *Surveys of the Territories* (**#725**). Portions reprinted in *To Improve the Public Surveys* (**#481**). See also **#267**.

Articles

262."The Cañons of the Colorado." *Scientific American Supplement* 5 (1878): 1869–1871, 1917–1919, 1965–1967; 6 (1878): 2075–2077, 2139–2141.

This series of articles, published April–August, consists of extracts from *Exploration*.

263."A Discourse on the Philosophy of the North American Indians." *Journal of the American Geographical Society of New York* 8 (1878): 251–268.

Same as *Outlines of the Philosophy of the North American Indians* (**#254**).

264."Geographical and Geological Survey of the Rocky Mountain Region." *Annual Report of the Board of Regents of the Smithsonian Institution ... 1877.* Washington: GPO, 1878. 67–87.

In the "Appendix to Report of the Secretary." Essentially the same as *Report on the Geographical and Geological Survey of the Rocky Mountain Region* (**#255**). See also **#258** and **#265**.

265."Geographical and Geological Survey of the Rocky Mountain Region Under the Direction of Professor J. W. Powell. Account of Work Performed During the Year 1877." *American Journal of Science and Arts* 3rd ser. 15 (1878): 342–358, 449–455.

Essentially the same as *Report on the Geographical and Geological Survey of the Rocky Mountain Region* (**#255**). See also **#258** and **#264**.

266."The Nationality of the Pueblos." *Rocky Mountain Presbyterian* 7.11 (1878): 1.

Brief article (1/2 column) that classifies the Pueblo people of New Mexico and Arizona into four language stocks.

267. "Report on the Methods of Surveying the Public Domain." *Survey of the Territories. Letter From the Acting President of the National Academy of Sciences Transmitting a Report on the Survey of the Territories.* 45th Cong., 3rd Sess., H. Misc. Doc. 5 Serial 1861. Washington: GPO, 1878. 14–37.

See **#261** for annotation. See also **#725**.

268. Statements. *Public Surveys: Testimony Before the Committee on the Public Lands.* 45th Cong., 2nd Sess., H. Misc. Doc. 55 Serial 1818. Washington: GPO, 1878. 1–6, 15, 32–38.

Powell proposes a change in the methods by which the public lands are surveyed and parceled. The land west of the 100th meridian should be surveyed according to its topography by the triangulation method. A scientific engineer, who would head the proposed Office of Surveyor-General, should direct surveying.

1879

Monographs

269. *Cost of Geographical Surveys.* 45th Cong., 3rd Sess., H. Ex. Doc. 72 Serial 1858. Washington: GPO, 1879.

Secretary Schurz transmits a brief report (6 pp.) from Powell endorsing the consolidation of the various geological surveys into one system, "in the interest of economy, and also manifestly in the interest of scientific research."

270. *Report on the Lands of the Arid Region of the United States: With a More Detailed Account of the Lands of Utah.* [2nd ed]. Washington: GPO, 1879.

Except for minor corrections, this is essentially the same as the 1878 edition (**#260**), which had gone out of print. Reprinted in **#574** and **#586**.

Articles

271. "Geographical and Geological Survey of the Rocky Mountain Region." *Annual Report of the Board of Regents of the Smithsonian Institution ... 1878.* Washington: GPO, 1879. 79–81.

In the "Appendix to the Report of the Secretary." Government explorations and surveys in 1878 extended to the Colorado Plateau south of the Grand Canyon. Four independent parties were in the field under the direction of G. K. Gilbert. George Wheeler surveyed an area in the western half of the region.

272. "Mythologic Philosophy." *Popular Science Monthly* 15 (1879): 795–808; 16 (1880): 856–866.

Slight revision of the paper Powell read as his vice-presidential address before the AAAS in 1879. Powell's view of mythology is presented within the context of the evolution of philosophy. The two "grand stages of philosophy" are the "mythologic" and the scientific. In mythologic philosophy, phenomena are explained by analogies derived from subjective human experiences. In scientific philosophy, phenomena are explained as orderly successions of events. Powell concludes, "When society shall have passed to complete integration in the unification of the nations, and differentiation is perfected in universal liberty, then the sole philosophy will be science."

Published by the AAAS in 1880 (**#277**, **#280**) and in the 1st report of the BAE in 1881 as "Sketch of the Mythology of the North American Indians" (**#290**, **#298**).

Abstracts, Remarks

273."Colorado River Country." *Mining and Scientific Press* 39 (1879): 223.

Transcription of Powell's address to the NAS on the Colorado River canyons and Pueblo Indians.

1880

Monographs

274.*Alphabet to Accompany Second Edition of "Introduction to the Study of Indian Languages.*" Washington: GPO, 1880.

Issued by the BAE with **#276**. See earlier edition (**#253**).

275.*Ethnology of North American Indian. Letter From the Secretary of the Smithsonian Institution, Recommending an Appropriation for Continuing the Ethnologic Researches Among the North American Indians.* 46th Cong., 2nd Sess., H. Misc. Doc. 35 Serial 1931. Washington: GPO, 1880.

Following a supportive letter from Secretary Baird, Powell requests $50,000 to continue ethnological research under the direction of the Smithsonian. He briefly relays the work of the BAE by category: somatology, philology, mythology, sociology, habits and customs, technology, archaeology, and history of Indian affairs.

276.*Introduction to the Study of Indian Languages, With Words, Phrases, and Sentences to Be Collected.* 2nd ed. Washington: GPO, 1880.

This revised and significantly enlarged edition of the 1877 publication (**#253**) contains Powell's revision of Whitney's alphabet; a section on houses and house life by Lewis Henry Morgan; a portion of a paper by J. Hammond Trumbull, "On the Best Method of Studying the North American Languages"; and four kinship charts by Morgan in the cover pocket. Powell's "The Rank of Indian Languages" (pp. 69c–74c) is largely the same as "On the Evolution of Language" (**#289**). Issued with *Alphabet* (**#274**). Reprinted in 1997 (**#590**).

277.*Mythologic Philosophy: An Address by John W. Powell (Vice President) Before the American Association for the Advancement of Science at Saratoga, N.Y., August 1879*. Salem, MA: Salem Pr, 1880.

Separate publication of the same in *Proceedings of the AAAS* (**#280**). First published in *Popular Science* 1879 (see **#272** for annotation).

Articles

278."Anthropology: Pueblo Indians." *American Naturalist* 14 (1880): 603–605.

A description of the four main languages of the Pueblo Indians.

279."Biographical Notice of Archibald Robertson Marvine." *Bulletin of the Philosophical Society of Washington, 1874–1878* 2 (1880): Appendix X, i–viii (53–60).

Marvine was a geologist who worked for the Hayden Survey. Powell praises Marvine's report on his work for its clear and simple language and presentation of complex geological structures.

Warman separate not seen. In *SMC* v. 20.

280."Mythologic Philosophy." *Proceedings of the American Association for the Advancement of Science* 28 (1880): 251–278.

See **#272** and **#277**.

281."Prefatory Note." *Report on the Geology of the High Plateaus of Utah*. By Clarence E. Dutton. Washington: GPO, 1880. vii–xiii.

In his introduction to this report of the Powell Survey (see **#117**), Powell describes the geology, geography, and topography of the Colorado Plateau. He notes that structural details may be found in the maps and graphs that accompany the report.

282."Sketch of Lewis H. Morgan, President of the American Association for the Advancement of Science." *Popular Science Monthly* 18 (1880): 114–121.

Powell praises Morgan's pioneering research among the North American Indians and his great work, *Ancient Society* (**#149**).

Remarks

283."The Lands of the Arid Region of the United States." *Bulletin of the Philosophical Society of Washington, 1874–1878* 2 (1880): 189.

Abstract of a "communication" from Powell at a meeting on 16 Mar. 1878, in which Powell states that less than 3% of Utah is arable.

In *SMC* v. 20.

284."Monoclinal Ridges." *Bulletin of the Philosophical Society of Washington, 1874–1878* 2 (1880): 74–76.

An abstract of a paper read at the 22 Apr. 1876 meeting of the PSW. Monoclinal ridges in the Rocky Mountains are composed of beds dipping in one direction. Faults affect the ridges in predictable ways. Monoclinal ridges are of two types: unequal degradation and displacement.

In *SMC* v. 20.

285."Poisons Among the North American Indians." *Bulletin of the Philosophical Society of Washington, 1874–1878* 2 (1880): 182–183.

Abstract of remarks made 22 Dec. 1877 regarding the use of deer liver for arrow tips. Powell also spoke of the arts of South American Indians are derived from a higher state of civilization than those of North American Indians.

In *SMC* v. 20.

286. Remarks. *Bulletin of the Philosophical Society of Washington, 1874–1878* 2 (1880): 44–45.

Remarks made 8 and 22 May 1875 in a discussion of the paper "On the Causes of the Glacial Period." Powell describes his observations on the distribution of gravel at sites where there is evidence of prehistoric human habitation.

In *SMC* v. 20.

287. Remarks. *Bulletin of the Philosophical Society of Washington, 1878–1880* 3 (1880): 137–139.

Powell's remarks follow a paper by J. Owen Dorsey, "On the Gentile System of the Omahas." Powell posits that social organization among

North American Indians had its origin in government by kinship, not patriarchy, citing the Wyandots as example.

In *SMC* v. 20.

1881

Monographs

288.*Abstract of Transactions of the Anthropological Society of Washington, D.C.: With the Annual Address of the President, for the First Year, Ending January 20, 1880, and for the Second Year, Ending January 18, 1881.* Washington: National Republican Printing House, 1881.

This publication was printed at the joint expense of Powell and the ASW. See **#195**.

In *SMC* v. 25.

289.*On the Evolution of Language, As Exhibited in the Specialization of the Grammatic Processes, the Differentiation of the Parts of Speech, and the Integration of the Sentence; From a Study of Indian Languages.* Washington: GPO, 1881.

The first annual address of the president of the ASW, delivered 2 Mar. 1880. Powell defines four "grammatic processes" and their purposes. The processes of change in language are then classed according to progress, which is measured by the degree of specialization and the organization and complexity of grammatical structure. Indian languages are found to be on the low end of evolution, compared to English, which "stands alone in the highest rank" in linguistic evolution though it "has but emerged from a barbaric condition" as a written language.

On cover: "Extracted from the first annual report of the Bureau of Ethnology" (see **#297**). The text appeared first as "The Rank of Indian Languages" in the second edition of *Introduction to the Study of Indian Languages* (**#276**). See also **#296**.

290.*Sketch of the Mythology of the North American Indians.* Washington: GPO, 1881.

On cover: "Extracted from the first annual report of the Bureau of Ethnology" (see **#298**). See **#272** for annotation.

291. *Wyandot Government: A Short Study of Tribal Society*. Washington: GPO, 1881.

Based on his work with a visiting Wyandot chief in Washington, this paper, read as an address to the AAAS (see **#301**) on 15 June 1880, focuses entirely on one tribe's social organization. Powell delineates the relations of kinship and marriage, on which the government of the Wyandot is based, and describes its various rules and regulations.

On cover: "Extracted from the first annual report of the Bureau of Ethnology" (**#300**). See **#299**.

292. *Wyandotte Government—A Short Study of Tribal Society; An Address*. Salem, MA: Salem Pr, 1881.

Pamphlet edition of address to the AAAS (**#301**). Published in the 1st annual report of the BAE (**#300**). See **#291** for annotation.

Articles

293. "[First Annual] Report [of the Director] of the Bureau of Ethnology." *First Annual Report of the Bureau of Ethnology to the Secretary of the Smithsonian Institution 1879–'80*. Washington: GPO, 1881. xi–xxxiii.

The first report of the BAE is dominated by Powell's work and his vision for the new agency. In his introduction, Powell predicts that objective ethnological studies of discoveries in North America will cause human history "to be rewritten, the whole philosophy of history reconstructed." After briefly reviewing the history of government ethnological work and its enabling legislation, Powell lays out his plan of organization for the work of the BAE. Scholars and specialists will be employed for research and the bureau will incite and organize research to be conducted by "collaborators at work throughout the country."

In addition to the director's report, there are four papers by Powell: "On the Evolution of Language" (**#297**); "Sketch of the Mythology of the North American Indians" (**#298**); "Wyandot Government: A Short Study of Tribal Society" (**#300**); and "On Limitations to the Use of Some Anthropologic Data" (**#295**).

Warman separate not seen.

294."On Limitations to the Use of Some Anthropologic Data." *Abstract of Transactions of the Anthropological Society of Washington* (1881): 113–136.

Read as the annual address of the president on 1 Feb. 1881. Powell contends that published reports of archaeological investigation in North America have been wrongly used to connect "the tribes of North America with peoples of so-called races of antiquity in other portions of the world." Objective, scientific research conducted in America overthrows these theories of "metaphysical writers." Anthropology needs individuals trained in scientific methods to establish anthropology as a science. Powell pioneers the use of the term "acculturation."

In *SMC* v. 25. Published in 1st annual report of the BAE (**#295**).

295."On Limitations to the Use of Some Anthropologic Data." *First Annual Report of the Bureau of Ethnology to the Secretary of the Smithsonian Institution 1879–'80*. Washington: GPO, 1881. 71–86.

See **#294**.

296."On the Evolution of Language, as Exhibited in the Specialization of the Grammatic Processes, the Differentiation of the Parts of Speech, and the Integration of the Sentence; From a Study of Indian Languages." *Abstract of Transactions of the Anthropological Society of Washington* (1881): 35–54.

See **#289** for annotation. Also published in 1st annual report of the BAE (**#297**).

297."On the Evolution of Language, as Exhibited in the Specialization of the Grammatic Processes, the Differentiation of the Parts of Speech, and the Integration of the Sentence; From a Study of Indian Languages. " *First Annual Report of the Bureau of Ethnology to the Secretary of the Smithsonian Institution 1879–'80*. Washington: GPO, 1881. 1–16.

See **#289** and **#296**.

298."Sketch of the Mythology of the North American Indians." *First Annual Report of the Bureau of Ethnology to the Secretary of the Smithsonian Institution 1879–'80*. Washington: GPO, 1881. 17–56.

See **#272** for annotation. See also **#290**.

299."Wyandot Government—A Short Study of Tribal Society." *Abstract of Transactions of the Anthropological Society of Washington* (1881): 76–92.

See **#291** for annotation. In *SMC* v. 25. See also **#300** and **#301**.

300."Wyandot Government: A Short Study of Tribal Society." *First Annual Report of the Bureau of Ethnology to the Secretary of the Smithsonian Institution 1879–'80*. Washington: GPO, 1881. 57–69.

See **#291** for annotation. Same as address to the AAAS (**#292**, **#301**).

301."Wyandotte Government. A Short Study of Tribal Society." *Proceedings of the American Association for the Advancement of Science, 29th Meeting at Boston, August 1880*. Salem, MA: AAAS, 1881. 675–688.

See **#291**, **#292**, and **#299**.

1882

Monographs

302.*Outlines of Sociology: Presidential Address to the Anthropological Society of Washington, D.C., February 2, 1882*. Washington, Judd & Detweiler, 1882.

Annual address of the ASW president. This also constituted one of the Saturday lectures at the USNM in April 1882. It is a compendium of Powell's views on the organization and evolution of the political state. Powell defines and classifies the state and its subsidiary units—government, law, and enforcement—and contrasts those of primitive societies with civilized ones. Government evolves in a society in order to establish and maintain justice among the differentiated classes and corporations. As societies become wealthy and complex, in order to deliver its prime function of securing the peace, government shifts its emphasis from kinship regulation to the business of regulating property. See also **#306** and **#307**.

303.*The Philosophic Bearings of Darwinism: An Address Delivered by John Wesley Powell Before the Biological Society of Washington at the Darwin Memorial Meeting, May 12, 1882*. Washington: Judd & Detweiler, 1882.

Pamphlet edition of "Darwin's Contributions to Philosophy" (**#313**).

Articles

304."Address of Major Powell." *Transactions of the American Institute of Mining Engineers* (1881–1882) 10 (1882): 232–236.

For mining engineers to be successful, they must have a sound basis in structural geology, which can "only be discovered and delineated by the great agencies of government." In the history of civilized states, technology has done more for mankind than sociology or philosophy. Philosophy is science, and technology is the father of science.

305."Introductory Address." *The Saturday Lectures Delivered in the Lecture-Room of the U.S. National Museum Under the Auspices of the Anthropological and Biological Societies of Washington, in March and April, 1882.* Washington: Judd & Detweiler, 1882. 1–3.

Powell inaugurates a series of popular science lectures that will present the scientific research of various federal agencies to the general public.

306."Outlines of Sociology." *Transactions of the Anthropological Society of Washington, 1879–1882* 1 (1882): 106–129.

See **#302** and **#307**.

307."Outlines of Sociology." *The Saturday Lectures Delivered in the Lecture-Room of the U. S. National Museum Under the Auspices of the Anthropological and Biological Societies of Washington, in March and April, 1882.* Washington: Judd & Detweiler, 1882. 60–82.

See **#302** and **#306**.

308."[Second Annual] Report of the Director of the United States Geological Survey." *Second Annual Report of the United States Geological Survey, 1880–'81.* Washington: GPO, 1882. i–lv.

Referring to Clarence King's resignation from the directorship of the USGS, Powell announces his own plan of operation, including a summary of studies and proposed publications. Powell then describes his system of uniform nomenclature and conventional characters for diagrams to be used in all government charts and atlases—the "American conventions."

Portions same as "Sur la nomenclature générale…" (**#310**). Warman separate not seen.

309."Statements." *Report of the Committee on Mines and Mining on Bills to Establish a Bureau of Mines and Mining and the Management of an Independent Organization Known As the U. S. Geological Survey.* 47th Cong., 1st Sess., H. Rept. 1065 Serial 2068. Washington: GPO, 1882. 4.

In testimony, Powell states that the work of the USGS is "1st. Topography; 2d. Structural Geology; 3d. Mining Geology" (Rabbitt, Mary C. *Minerals, Lands, and Geology for the Common Defence and General Welfare, V. 2: 1879–1904*, 1980, p. 65). Not seen.

310."Sur la nomenclature générale, sur la coloriage et les signes conventionnels des Cartes géologiques." *Congrès Géologique Internationale, Compte Rendu de la 2me Session, Bologne, 1881*. Bologna: Fava et Garagnani, 1882. 627–642.

Powell transmits to the International Congress of Geologists this paper on geological nomenclature, a color scheme for geological charts, and a chart of symbols for geological sections as adopted for use by the USGS. See also **#372**.

1883

Monographs

311.*Human Evolution; Presidential Address Delivered Before the Anthropological Society of Washington, November 6, 1883*. Washington: Judd & Detweiler, 1883.

Powell identifies various types of data through which the history of man's evolution is discovered. The earliest institutions of mankind were simpler in structure and less numerous in differentiation in earlier stages than later civilizations. All human activities can be divided into five categories—arts, industry, institutions, languages, and philosophy—upon which the evolution of the human mind depends and proceeds. He then traces the course of evolution from the lower to the higher in each of the five categories. Mankind is superior to lower animals not by the doctrine of survival of the fittest, but because man can transfigure selfishness into love. Published also in *Transactions of the ASW* (**#314**).

312.*The Three Methods of Evolution*. Washington: Judd & Detweiler, 1883.

Annual address of the president of the PSW, 8 Dec. 1883. Powell proposes three "grand classes of phenomena, constituting three kingdoms of matter and representing three methods of evolution" that in turn correspond to three great classes of combinations. Evolution, which is systematic progressive change, takes place by three methods: direct adaptation to the environment, indirect adaptation, and adaptation of the environment to man. Man is superior to animals because he is the master of his own destiny and not a creature of his environment. Among notable features of this essay are Powell's invented vocabulary, florid metaphors, and fixation on numerical symmetry, all of which become more exaggerated in succeeding years. Published in the *Bulletin of the PSW* (**#329**).

Articles

313."Darwin's Contributions to Philosophy." *Proceedings of the Biological Society of Washington, 1880–1882* 1 (1883): 60–70.

> Presented as an address on the occasion of the Darwin Memorial Meeting of the BSW on 12 May 1882. Powell classifies philosophies of the world into three groups: mythologic, metaphysic, and scientific. Metaphysics is formal logic and serves as a "stepping-stone" between mythology (superficial analogy) and science (induction and hypothesis). Powell praises Darwin "as a light in the scientific world" for hypothesizing his theory of evolution, thus demonstrating that metaphysical philosophy is false.
>
> Published as a pamphlet with new title, *The Philosophic Bearings of Darwinism* (**#303**). In *SMC* v. 25.

314."Human Evolution." *Transactions of the Anthropological Society of Washington, 1882–1883* 2 (1883): 176–208.

> See **#311** for annotation.

315. Letters. *Report on the Sundry Civil Expenses Bill.* 47th Cong., 2nd Sess., H. Rept. 1960 Serial 2160. Washington: GPO, 1883.

> "Includes three letters from Powell on the salaries of chief geologists, an explanation of estimates for appropriations, and a statement on the district over which the Survey was extending its operations" (**#859**, p. 377). Not seen.

316."Second Annual Report [of the Director] of the Bureau of Ethnology." *Second Annual Report of the Bureau of Ethnology to the Secretary of the Smithsonian Institution 1880–'81.* Washington: GPO, 1883. xv–xxxvii.

> Powell reiterates the BAE's plan of operation: direct employment of scholars and specialists, combined with "inciting and guiding research immediately conducted by collaborators not directly connected with the Bureau." Describes improvements to the second edition of *Introduction to the Study of Indian Languages* (**#276**).
>
> Warman separate not seen.

317."[Third Annual] Report of the Director of the United States Geological Survey." *Third Annual Report of the United States Geological Survey, 1881–'82.* 1883. iii–xviii.

> The USGS will build on the work of predecessor surveys in the Colorado Plateau region by preparing a topographic map of the region.
>
> Warman separate not seen.

318."Ward's *Dynamic Sociology*." *Science* 2 (1883): 45–49, 105–108, 171–174, 222–226.

> There are four parts to Powell's review of Lester Ward's classic two-volume work (**#152**). Part I summarizes and identifies key tenets. Part II elaborates on Ward's view that social progress is due to the struggle for happiness and the welfare of all, rather than survival of the fittest. Thus human evolution is different from biological evolution and in direct opposition to the views of Herbert Spencer. In Part III, Powell criticizes Ward's classification of phenomena for being "philosophically narrow." Powell would add a third group called "anthropology." The last segment, Part IV, compares and contrasts Ward and Lewis Morgan. Powell states his own views on social organization, which projects a final evolutionary stage wherein languages, philosophies, nations, and arts will be united by the philosophy of science, "the truth."

Abstracts, Remarks

319."A Classification of the Sciences." *Science* 2 (1883): 370.

> Abstract of a paper read at the Minneapolis meeting of the AAAS in August 1883. The sciences are classified according to Powell's scheme into three groups: physics, biology, and anthropology. An article with the same title in the *American Anthropologist* (**#546**) is quite different.

320."Informal Remarks on Moraines and Terraces." With J. W. Dawson. *Science* 2 (1883): 321.

> Summary of remarks by Powell and Dawson at the Minneapolis meeting of the AAAS in August 1883. They maintains that a large number of different agencies, often within the same geographical area, are responsible for the feature known as terrace.

1884

Monographs

321. *Address by Hon. John W. Powell, LL.D., Delivered at the Inauguration of the Corcoran School of Science and Art, in the Columbian University, Washington, D.C., October 1, 1884.* Washington: Gibson Bros. Printers, 1884.

> Powell reinterprets the growth of civilization in order to demonstrate that modern civilization, based on the principles of science, is superior to all others. In upholding the ideals of a modern scientific education, the establishment of the Corcoran School is a "landmark in the progress of culture and the history of education." In pursuing a scientific education, students will gain aesthetic training, integrity, and charity. The second half is published as "The Larger Import of Scientific Education" (**#341**).

Articles

322. "Certain Principles of Primitive Law." *Science* 4 (1884): 436–437.

> First of a two-part series (see **#325**). Powell defines law as a rule of conduct enforced by an organized society and then establishes three principles of primitive law. "Savage" law is not concerned with justice but with preventing and terminating controversy. Powell demonstrates these principles with examples of laws found among North American Indians. Also published in the 3rd annual report of the BAE (**#328**).

323. "[Fourth Annual] Report of the Director of the United States Geological Survey." *Fourth Annual Report of the United States Geological Survey, 1882–'83*. Washington: GPO, 1884. iii–xxxii.

> Powell announces that the congressional act making appropriations for the USGS for 1882–1883 requires the preparation of a geologic map of the United States, which in turn necessitates the preparation of a topographic map. He summarizes the topographic surveys conducted by the western surveys, the Coast and Geodetic Survey, and various state surveys.
>
> Warman separate not seen.

324. "Marriage Law in Savagery." *Science* 4 (1884): 471–473.

> This paper was presented to the NAS in November 1883 and titled "Marriage Institutions in Tribal Society." Powell states that among all

"savage" societies, marriage selection is by legal appointment and not by personal choice, so that controversy may be avoided. However, legal appointment may be modified by three forms of marriage that allow varying degrees of personal choice. The terms "exogamy" and "endogamy" are misleading and should be discarded. Second of a two-part series (see **#321**). Also published in the 3rd annual report of the BAE (**#328**).

325."On Activital Similarities." *Third Annual Report of the Bureau of Ethnology to the Secretary of the Smithsonian Institution, 1881–'82*. Washington: GPO, 1884. lxv–lxxiv.

Powell states that there are diverse means for accomplishing the same ends in four areas of human activity: arts, institutions, languages, and opinions. Functional activities that are similar may be classified in one of two ways. Autogenous similarities develop independently among different peoples in different places. Syngenous activities have a common origin. Powell delineates nine principles for scientific interpretation of such activities. See also **#328**.

326."On the Fundamental Theory of Dynamic Geology." *Science* 3 (1884): 511–513.

Continuing the argument presented in "On the State of the Interior of the Earth" (**#327**), Powell postulates that the crust of the earth is a solid, thin shell around a fluid interior and is "in a state of floating equilibrium." The thickness of the crust varies, but evidence points to its becoming more heterogeneous with time.

327. "On the State of the Interior of the Earth." *Science* 3 (1884): 480–482.

Using inductive reasoning, based largely on his observations in Utah and Arizona, Powell hypothesizes that the interior of the earth is fluid rather than solid. Followed by **#326**.

328. "Third Annual Report [of the Director] of the Bureau of Ethnology." *Third Annual Report of the Bureau of Ethnology to the Secretary of the Smithsonian Institution, 1881–'82*. Washington: GPO, 1884. xiii–lxxiv.

In addition to "On Actival Similarities" (**#325**), this report includes three essays embedded in Powell's comments on an accompanying paper by J. Owen Dorsey, "Omaha Sociology.". "Kinship and the Tribe," "Kinship and the Clan," and "Tribal Marriage Law" (xxxviii–lxii) describe tribal kinship and explain the kinship system and marriage laws of clans constituting a tribe. "Kinship and the Tribe" is largely the same as "On the Organization of the Tribe" (**#344**). "Tribal

Marriage Law" is excerpted in "Certain Principles" (**#322**) and "Marriage Law in Savagery" (**#324**).

Warman separate not seen.

329."The Three Methods of Evolution." *Bulletin of the Philosophical Society of Washington, 1883* 6 (1884). xxvii–lii.

See **#312** for annotation. In *SMC* v. 33.

330."United States Geological Survey." *Annual Report of the Board of Regents of the Smithsonian Institution ... 1882*. Washington: GPO, 1884. 47–49.

Largely the same as the Director's Report in the USGS annual report for the year (see **#317**).

Abstracts, Remarks

331."The Cañons of the Colorado." *Chautauquan* 4 (1884): 564–566.

Lengthy summary of a Saturday lecture (**#305**) delivered by Powell on 26 Apr. 1884 in the USNM—a dramatic rendition of the 1869 expedition.

1885

Monographs

332.*From Savagery to Barbarism; Presidential Address to the Anthropological Society of Washington, D.C.* Washington: Judd & Detweiler, 1885.

Annual address, delivered 3 Feb. 1885. Here and in "From Barbarism to Civilization" (**#362**), Powell further develops his theory of the evolution of mankind. The course of human culture is found in the evolution of the humanities: arts, institutions, languages, opinions, and intellections. Powell describes the savage and barbaric stages of these five "great classes of human activities." The transition between the stages is marked by the invention of agriculture and the domestication of animals. Powell concludes by distinguishing the evolution of mankind from biological evolution, which in his view are absolutely unrelated. Man has progressed by his endeavor to secure happiness through human activities, not by survival of the fittest. Rather, man has come to work directly for others and indirectly for himself: "So are we all—servants to many masters and masters of many servants."

Pamphlet edition of article in *Transactions of the ASW* (**#338**).

333. *On the Organization of Scientific Work of the General Government.* Washington: GPO, 1885.

Extracts from *Testimony Before the Joint Commission* (**#843**, pp. 1–49, 162–180, 184–209, 378–451, 460–468). See **#356** for annotation. See also expanded 1886 edition (**#349**) and "The Administration of the Scientific Work of the General Government" (**#334**).

Articles

334. "The Administration of the Scientific Work of the General Government." *Science* 5 (1885): 51–55.

Powell's testimony and prepared statements presented before the Allison Commission on 5 Dec. 1884. Powell outlines his philosophy of the administration of scientific agencies. They should be centrally managed, eliminating duplication and competition, and the work of organization should originate with experts who are themselves engaged in the investigation. See **#333**, **#356**, and **#843**.

335. "Bureau of Ethnology [and] United States Geological Survey. *Annual Report of the Board of Regents of the Smithsonian Institution ... 1883.* 1885. 56–81.

Largely the same as the Director's Report in the BAE and USGS annual reports for the year (see **#323**, **#351**).

336. "Bureau of Ethnology [and] United States Geological Survey [Accounts of Their Operations]." *Annual Report of the Board of Regents of the Smithsonian Institution ... 1884.* 1885. 67–95.

Largely the same as the Director's Report in the BAE and USGS annual reports for the year (see **#337**, **#358**).

337. "[Fifth Annual] Report of the Director of the United States Geological Survey." *Fifth Annual Report of the United States Geological Survey, 1883–'84.* Washington: GPO, 1885. iii–xxxvi.

Powell and WJ McGee report that a preliminary geological map of the nation has been completed.

Warman separate not seen.

338. "From Savagery to Barbarism." *Transactions of the Anthropological Society of Washington, 1883–1885* 3 (1885): 173–196.

See **#332** for annotation.

339."The Indians Are the Mound-Builders." *Science* 5 (1885): 267.

Response to a review of *Prehistoric America*, by Marquis de Nadaillac. Powell agrees that the mound builders are North American Indian tribes, but disagrees that these builders constitute a single distinct race.

340."Inheritance Among the Ancient Arabs." *Science* 5 (1885): 16–18.

Review of *Das Matriarchat (das Muterrecht) bei den alten Arabern*, by G. A. Wilken. Powell agrees that "uterine descent" once existed among the ancient Arabs, but criticizes the author's misconceptions of communal marriage, exogamy and endogamy, and patriarchy and matriarchy.

341."The Larger Import of Scientific Education." *Popular Science Monthly* 26 (1885): 452–456.

Same as the second half of Powell's address delivered at the Corcoran School of Science and Arts (**#321**).

342."Letter From the Director of the Geological Survey." *Report of the National Academy of Sciences for the Year 1884*. Washington: GPO, 1885. 48–50.

The letter is dated 8 Sept. 1884 and addressed to General M. C. Meigs, who chaired the NAS committee convened at the request of the Joint Commission to Consider the Present Organization of the Signal Service, Geological Survey, Coast and Geodetic Survey, and the Hydrographic Office of the Navy Department, also known as the Allison Commission. The Commission's task was to consider ways to end waste and duplication in the federal government. The committee asked the agency directors to submit letters describing their operations. Powell's reply on the operations and work of the USGS stresses topographic maps and geological surveys. The various divisions and laboratories within the survey are outlined.

Reprinted in *Testimony Before the Joint Commission ...*" (**#843**, pp. 26*–27*). See also **#347**, **#355**, and **#356**.

343."Note." *Proof-Sheets of a Bibliography of the Language of the North American Indians*. James Constantine Pilling. Washington: GPO, 1885. 3.

Powell comments on the extensive bibliography of Indian languages accumulated by the Smithsonian and the BAE, and compliments Pilling on his hard work. Only 110 copies of this publication were printed for distribution to BAE collaborators.

344."On the Organization of the Tribe." *Scientific American Supplement* 19 (1885): 7889–7891.

Read before the NAS on 22 Apr. 1885, this paper explains the fundamental principles of tribal kinship, including several characteristics of the clan. Except for its conclusion, this is the same as "Kinship and the Tribe" and "Kinship and the Clan" (**#328**).

345."The Organization and Plan of the United States Geological Survey." *American Journal of Science* 3rd ser. 29 (1885): 93–102.

Presented to the NAS in October 1884. Powell emphasizes the topographic maps and geological survey work of the USGS plan for a topographic survey of the United States. He explains its operations, organizational divisions, personnel classifications, and finances. This is an elaboration of his letter in the *Report of the NAS for 1884* (**#342**). See also **#355**.

Warman separate not seen.

346."*The Patriarchal Theory*." *Science* 5 (1885): 345–348.

A review of *The Patriarchal Theory, Based on the Papers of the Late John Ferguson McLennan* (1885). While McLennan was correct in identifying the patriarchal society of the Romans as an exception among ancient tribes, he erred in his theories relating to primitive marriage and kinship. Powell concludes with a call for a "piece of good destructive criticism" of Herbert Spencer's theory of ancestral worship.

1886

Monographs

347._The Geological Survey._ Washington: GPO, 1886.

This is a transcript, published as a pamphlet, of Powell's statement on 13 May 1886 before the Allison Commission in response to Bill H.R. 8320, introduced by Hilary Herbert to cut appropriations for printing the publications of the USGS. Powell provides data showing that the cost of publication is relatively inexpensive compared to publications by state surveys and defends the printing of topographic maps. Powell puts Herbert on the spot for an earlier statement in which Herbert implied that Powell's employment practices were corrupt.

See also **#357** and **#831**.

348.*Letter From the Secretary of the Interior, Transmitting Report of the Director of the U.S. Geological Survey on the Merits of the Bill S. 1907 "to Facilitate the Settlement and Development of the Resources of the Territory of Alaska, and Open an Overland and Commercial Route Between the United States, Asiatic Russia, and Japan," and the Feasibility of the Construction of the Railroad Proposed.*" 49th Cong., 2nd Sess., S. Misc. Doc. 22 Serial 2450. Washington: GPO, 1886.

> In response to a request from Congress to assess the feasibility of constructing a railroad from the northwest United States to the Bering Sea, Powell gives reasons for his conclusion that a line is possible but the best location requires "extended and expensive surveys."

349.*On the Organization of Scientific Work of the General Government.* Washington: GPO, 1886.

> This is the same as the 1885 publication (**#333**) with the addition of "Part 2—Additional Statements" from *Testimony Before the Joint Commission* (**#843**). Contents include "Statement Traversing..." (**#357**). See also **#334**.

Articles

350. "The Cause of Earthquakes." *Forum* 2 (1886): 370–391.

> Written in response to newspaper articles that offer somewhat mystical explanations for a destructive earthquake in Charleston, South Carolina. Powell offers a scientific view of the event for the lay audience. The surface of the earth is described as a series of formations that are a kind of structural shell floating on the earth's core. He explains earthquakes in terms of geological displacement, drawing heavily upon the Colorado River country for examples.

351."Fourth Annual Report [of the Director] of the Bureau of Ethnology." *Fourth Annual Report of the Bureau of Ethnology to the Secretary of the Smithsonian Institution 1882–'83.* Washington: GPO, 1886. xxvii–lxiii.

> Powell reports on the directive from Congress that the BAE add archaeological investigations of the eastern United States to its agenda. In response, a Division of Mound Explorations is organized in 1882 under the direction of Cyrus Thomas. Includes a chart of the BAE's scheme of conventions for "Archaeologic Cartography."
>
> Warman separate not seen.

352. Letter. *The Work of the International Congress of Geologists, and of Its Committees, Published by the American Committee.* Berlin: International Congress of Geologists, American Committee, 1886. 106–108.

The letter, dated 26 Mar. 1886, is addressed to Persifor Frazer. Powell criticizes the European map of the general system of cartography as "incompetent" and commends the American Committee and the American system. Appears as "Note 2 ... received after the pamphlet had been made up."

353. Review. "Conn's *Evolution of To-Day.*" *Science* 8 (1886): 264–265.

A book review. Powell thinks the work is unsatisfactory, chiefly because it is more speculative than scientific. H. W. Conn refers primarily to human physical evolution rather than "those characteristics which make man man—the growth of human activities."

354. "The Scientific Bureaus [sic] Bill to Restrict the Geological Survey; Director Powell's Criticisms on the Bill of the Joint Commission." *New York Daily Tribune.* 3 May 1886.

In response to a bill by Representative Herbert "restricting the work of the Geological Survey," Powell points out inconsistencies and errors in four key sections of the bill. The bill would prohibit expenditure of money by the USGS for work in paleontology and require the survey to sell laboratories no longer needed "after passage of this act." Publications of USGS work would become the property of the employees, essentially requiring self-publication, which Powell observes would result in work performed only by men of wealth.

355."[Sixth Annual] Report of the Director of the United States Geological Survey." *Sixth Annual Report of the United States Geological Survey, 1884–'85.* Washington: GPO, 1886. iii–xxix.

In this and his "Seventh Annual Report" (**#377**), Powell sets forth a plan for a topographic survey of the United States. The maps are to be used for a variety of practical and civil (not military) purposes. He further outlines a general plan for the work of the USGS and explains its operations, organizational divisions, and personnel classifications. Includes Powell's 1884 letter to General Meigs (**#342**). See also **#345** and **#843**.

Warman separate not seen.

356."Statement of the Director of the Geological Survey." *Testimony Before the Joint Commission to Consider the Present Organizations of the Signal Service, Geological Survey, Coast and Geodetic Survey and the Hydrographic Office of* *the Navy Department, With a View to Secure Greater Efficiency and Economy of Administration of the Public Service of Said Bureaus.* 49th Cong., 1st Sess., S. Misc. Doc. 82 Serial 2345. Washington: GPO, 1886. (1–49, 162–180, 184–209, 378–345, 460–468, 637–700, 896–912, 1019–1045, 1070–1084). [Rpt. New York: Arno Pr, 1980.]

In his testimony (see **#843**), Powell is called upon to testify in person regarding the work of the USGS, distinguishing its work from the other agencies. He responds by describing the survey's topographic maps and touting their practical benefits. Although the NAS had recommended the establishment of a Dept. of Science, Powell contends that each agency should remain independent in order to freely pursue research. Detailed accounts of the USGS's expenses, including salaries and rations for its employees, are submitted. Finally, in a response to a letter by Alexander Agassiz condemning the survey for waste and patronage, Powell defends his agency, arguing that government research promotes private research and stimulates the economic well being of the nation. The majority report vindicated Powell's position, but the minority report by Congressman Henry Herbert included a bill to curtail survey publications (see **#357**).

Powell's letter to the NAS is reprinted in prefatory material on pages 26*–27* (see **#342**). See also **#333**, **#334**, **#345**, and **#349**.

357. Statement ["Traversing Certain Averments of House Report No. 2214"]. ... *Limiting the Printing and Engraving for the Geological Survey....* 49th Cong., 1st Sess., H. Rept. 2740 Serial 2443. Washington: GPO, 1886. 98–125.

Representative Herbert of the Allison Commission presents a minority bill to stop the publications of all federal surveys (see **#831**). Powell had filed a letter of rebuttal to the minority report, which is excerpted here with Herbert's comments. Powell defends in detail the cost and utility of the publications of the USGS. Powell counters Herbert's accusation that the survey threatened to monopolize research by identifying the many agencies involved in geological research and states that the integrity of scientists would prevent any such monopoly. See also **#347** and **#349**.

1887

Articles

358."Fifth Annual Report [of the Director] of the Bureau of Ethnology." *Fifth Annual Report of the Bureau of Ethnology to the Secretary of the Smithsonian Institution 1883–'84*. Washington: GPO, 1887. xvii–liii.

Powell prefaces a report by Cyrus Thomas on mound explorations by concluding that all mounds in the northeastern region thus far examined can be attributed to the tribes currently inhabiting the region and their ancestors.

Warman separate not seen.

359."Museums of Ethnology and Their Classification." *Science* 9 (1887): 612–614.

Written in response to criticism by Franz Boas of an exhibit at the USNM organized by Otis Mason. Powell defends the exhibit, in which artifacts were arranged according to their rank on the evolutionary scale of human activities. Boas preferred a geographic distribution based on ethnic similarities, whereas Powell's underlying principle of classification is the unity of mankind.

1888

Monographs

360.*Competition as a Factor in Human Evolution; a Paper Read Before the Anthropological Society of Washington*. Washington: n.p., 1888.

Pamphlet edition of article from *American Anthropologist* (see **#365**). Continuing his thesis in "From Savagery to Barbarism" (**#338**) and "From Barbarism to Civilization" (**#362**), Powell argues that humans evolve not by natural selection but human selection. Society advances by both competition and cooperation, as well as by man's own invention and development of the arts. Powell takes a surprising philosophical and political turn by tackling the issue of the division between capital and labor. Economic and social justice will occur when the "professional classes ... learn this lesson from history, that when wrongs arise in any class of society those wrongs must ultimately be righted." Powell expresses his "towering faith" in his fellow man. Reprinted in *Seeing Things Whole* (**#213**).

361.*Eulogy of Professor Baird; An Address Delivered at a Meeting Held January 11, 1888, Under the Joint Auspices of the Anthropological, Biological and Philosophical Societies of Washington.* Washington: Judd & Detweiler, 1888.

Baird's superior qualities are noted in three respects: as an organizer of the agencies of research, as a man of kindness and modesty in all matters, and as a director of the work of research. This was published also as "The Personal Characteristics of Professor Baird" (**#373**) and "Address of Major Powell..." (**#363**). See also **#374** and **#408**.

362.*From Barbarism to Civilization.* Washington: n.p., 1888.

Delivered 16 Mar. 1886 as the annual address of the retiring president of the ASW, this work continues "From Savagery to Barbarism" (**#332**). First, Powell identifies errors in the current literature of anthropology, carefully emphasizing that savagery, barbarism, and civilization are stages of culture, not characteristics of individuals. Powell then traces the transition of the "five great classes of human activities" from the barbaric to the civilized. The conclusion is a series of definitions that condenses the characteristics of each stage into single sentence constructs. This is a pamphlet edition of an article *American Anthropologist* (**#368**). See also **#366**. Reprinted in *Seeing Things Whole* (**#213**).

Articles

363."Address of Major Powell in Memory of Professor Baird." *Science Supplement* 11 (1888): 25–26.

See **#361** for annotation. See also **#373**, **#374**, and **#408**.

364."Communication on the American Report of the International Congress of Geologists." *American Journal of Science* 3rd ser. 36 (1888): 476a–476e.

Letter to the editor of the journal in which Powell protests the attachment of his name to the American report for the International Congress of Geologists in London because he is an author of none of the papers. Powell cannot agree with the proposed system of geological conventions. He provides an abstract of his own report to the committee on quaternary formations and the need for a cartographic scheme for such.

Warman separate not seen.

365."Competition as a Factor in Human Evolution." *American Anthropologist* old ser. 1 (1888): 297–323.

See **#360** for annotation. Reprinted in *Seeing Things Whole* (**#213**).

366."The Course of Human Progress." *Science* 11 (1888): 220–222.

Extract of a lecture delivered in the USNM on 5 May 1888 as one in a series of free lectures under the auspices of the PSW, BSW, and ASW. Text is essentially the same as "From Barbarism to Civilization" (**#362**).

367."Evolution in Civilized Man." *Science* 11 (1888): 112–116.

Lengthy extract from "Competition as a Factor in Human Evolution" (**#365**).

368."From Barbarism to Civilization." *American Anthropologist* old ser. 1 (1888): 97–123.

See pamphlet **#362** for annotation. See also **#366**.

369."The Geological Survey; Statement of J. W. Powell." *Statements Before Subcommittee on the Sundry Civil Bill.* 50th Cong., 1st Sess., S. Rept. 1814 Serial 2525. Washington: GPO, 1888. 63–72, 111–126.

The Irrigation Survey was not officially instituted until the Sundry Civil Expenses Bill passed 2 Oct. 1888, appropriating $100,000 for its operation. In hearings before a subcommittee of the Senate Committee on Appropriations, Powell submits a prepared statement requesting additional funding for the Irrigation Survey and a map indicating where irrigation might be used for agriculture. During questioning, he explains that a topographical map is the first step in identifying potential sites for the construction of reservoirs. He goes on to discuss in detail his intent for the survey: that irrigation districts will be established and the people residing within it would have the authority to contract with capitalists to provide water to other users for a fee. The statement appears also in the 10th annual report of the USGS (**#416**).

370."The Laws of Hydraulic Degradation." *Mining Industry* 3 (1888): 226–228, 234–236.

A technical geological paper read before the NAS in New Haven in November 1888. The surface of the earth is degraded by three methods of hydraulic degradation: erosion, corrasion, and sapping. The forces of degradation are established by nature, which man can only control to a limited degree. Five methods of engineering are presented, each of which can control rivers for the protection of floodplains. Also in *Science* (**#371**).

371."The Laws of Hydraulic Degradation." *Science* 12 (1888): 229–233.

Same as **#370**.

372."Methods of Geologic Cartography in Use by the United States Geological Survey." *Congrès Géologique Internationale, Compte Rendu de la 3me Session, Berlin, 1885*. Berlin: A. W. Schade's Buchdruckerei, 1888. 221–240.

WJ McGee presented this paper to the International Congress of Geologists on behalf of Powell, who did not attend. Powell details modifications made by the USGS to its cartographic conventions and color scheme since his first paper on this topic was delivered to the Congress in 1881 (see **#310**).

Warman separate not seen.

373."The Personal Characteristics of Professor Baird. " *Bulletin of the Philosophical Society of Washington, 1887* 10 (1888): 71–77.

Published also as *Eulogy of Professor Baird* (**#361**) and "Address of Major Powell in Memory of Professor Baird" (**#363**). See also **#374** and **#408**.

In *SMC* v. 33.

374."The Personal Characteristics of Professor Baird." *Proceedings at a Meeting Commemorative of the Life and Scientific Work of Spencer Fullerton Baird, Held Jan. 11, 1888, Under the Joint Auspices of the Anthropological, Biological and Philosophical Societies of Washington.* Washington: Judd & Detweiler, 1888. 31–37.

See **#361**, **#363**, **#373**, and **#408**.

375.["Report Concerning the Irrigation of Certain Lands"]. *Letter From the Secretary of the Interior, Transmitting, in Response to Senate Resolution of Feb. 13, 1888, Report Concerning the Irrigation of Certain Lands.* 50th Cong., 1st Sess., S. Ex. Doc. 134 Serial 2513. Washington: GPO, 1888. 3–6.

Powell states that the conclusions reached in *Report on the Lands* are still valid. By extending the work of the USGS over the arid region, local surveys can be combined with information from existing topographic work to identify sites for reservoirs and other irrigation uses.

376.["Report Relative to the Reservoirs for the Storage of Water in the Arid Regions of the United States"]. *Letter From the Secretary of the Interior, Transmitting, in Response to Senate Resolution of March 27, 1888, Report Relative to the Reservoirs for the Storage of Water in the Arid Regions of the United States.* 50th Cong., 1st Sess., S. Ex. Doc. 163 Serial 2513. Washington: GPO, 1888. 2–6.

> Powell increases his estimate of the percentage of arid land that would be redeemable by irrigation. He considers factors that are entailed in surveying the land for the selection of sites for water storage and recommends an appropriation of $250,000 for the first year of surveying by the USGS.

377."Seventh Annual Report [of the Director] of the United States Geological Survey." *Seventh Annual Report of the United States Geological Survey, 1885–'86.* Washington: GPO, 1888. iii–xx, 2–42.

> Powell continues his report in the 6th annual report of the USGS (**#355**) on the organization of the USGS and plans for the topographic survey. See also "Statement of the Organization …" (**#379**).
>
> Warman separate not seen.

378."Sixth Annual Report [of the Director] of the Bureau of Ethnology." *Sixth Annual Report of the Bureau of Ethnology to the Secretary of the Smithsonian Institution 1884–'85.* Washington: GPO, 1888. xxiii–lviii.

> In the section "Office Work," Powell announces that work on a synonymy of the tribes of North America has significantly progressed. The BAE is in the process of adopting a code of rules for nomenclature.
>
> Warman separate not seen.

379. "Statement of the Organization, Business Methods and Work of the United States Geological Survey." *Report of the Select Committee of the United States Senate ... to Inquire Into and Examine the Methods of Business and Work in the Executive Departments....* 50th Cong, 1st Sess., S. Rept. 507, Pt. 2 Serial 2522. Washington: GPO, 1888. 377–409.

> This detailed, comprehensive report on the operation of the USGS is largely the same as Powell's director's reports in the 7th and 8th annual reports of the USGS (see **#377**, **#387**). Includes a table of contents and an impressive "list of exhibits" that includes everything from address slips to a "record of latitudes."
>
> Warman separate not seen.

380."The Three Americas Permanent Exhibition." *Science* 11 (1888): 303–305.

Lengthy extracts from Powell's address before the House Committee on Commerce regarding a proposal to establish an exhibition in Washington in 1892 to celebrate the 400th anniversary of the discovery of America by Columbus. Powell speaks of Columbus as "a great scholar, a scientific investigator, a man whose insight penetrated to the great secrets of nature in the light of the science of his time … [and] with a sublime faith in scientific inductions he sailed into an unknown sea." Powell eloquently laments the loss of opportunity to reconstruct the ancient history of America as "every dust-laden breeze buries them deeper" and expresses hope that the anniversary of the discovery of America will result in a gathering of materials of archaeology, which will be deposited in a new "grand international museum."

381."Trees on Arid Lands." *Science* 12 (1888): 170–171.

Powell argues against the popular fiction that tree planting can alter the arid climate of the West. However, the agricultural climate "may be materially modified" by planting trees adapted to aridity, irrigating, and protecting forests from fires. In this way, man can control his environment by using the laws of nature for his own purpose. In *Seeing Things Whole* (**#213**).

Abstracts, Remarks

382."Major Powell's Report; Operations of the National Survey During the Year Ending June 30, 1888." *Science* 12 (1888): 148–150.

Summary and extracts from the USGS 9th annual report (**#392**).

383."Physical Features of the Atlantic Coast." *Johns Hopkins University Circulars* 7.65 (1888): 73.

Abstract of a lecture on 5 Jan. 1888 in Hopkins Hall, Johns Hopkins University. The country bordering the Atlantic coast consists of two physical provinces between the Appalachians and the Atlantic. They are the Piedmont Region and the Plain of the Coast.

1889

Monographs

384.*Evolution of Music From Dance to Symphony.* Salem, MA: Salem Pr, 1889.

Read for Powell by G. K. Gilbert at the 38th meeting of the AAAS in Toronto, August 1889. Characterized by W. M. Davis as "rhapsodic" (see **#611**), this address traces the development of music through four stages, from rhythm, to melody, to harmony, then symphony. From its origin in primitive dance to the works of Wagner, who has "killed the jailers" of mathematical dogmas, music exemplifies human evolution through its three great stages of cultural progress. The fourth stage in this progression is enlightenment.

This is an offprint from the *Proceedings of the AAAS* (see **#388**) and printed at the head of the title is: "Address by J. W. Powell, President of the American Association for the Advancement of Science, delivered at the Toronto meeting, August, 1889."

Articles

385."Bureau of the Ethnology [and] United States Geological Survey." *Annual Report of the Board of Regents of the Smithsonian Institution…1886.* Vol. 1. Washington: GPO, 1889. 51–80.

Largely the same as the Director's Report in the annual reports of the BAE and USGS for the year (see **#377**, **#427**).

386."Bureau of Ethnology." *Annual Report of the Board of Regents of the Smithsonian Institution…1887.* Vol. 1. Washington: GPO, 1889. 20–27.

Largely the same as the Director's Report in the BAE's annual report for the year (see **#418**).

387."Eighth Annual Report of the Director of the United States Geological Survey." *Eighth Annual Report of the United States Geological Survey, 1886–'87.* Pt. I. Washington: GPO, 1889. iii–xix, 1–93.

Powell presents a comprehensive, detailed statement of the business organization of the USGS, which is comprised of these systems: fiscal, custodial, museum, illustration, editorial, document, library, stationery, and correspondence. Continued from the 7th annual report of the USGS (**#377**). See also **#379**.

Warman separate not seen.

388."Evolution of Music From Dance to Symphony." *Proceedings of the American Association for the Advancement of Science* 38 (1889). 1–21.

See **#384** for annotation.

389."Irrigation of Arid Lands." *Sundry Civil Appropriation Bill.* 50th Cong., 2nd Sess., S. Rept. 2613 Serial 2619. Washington: GPO, 1889. 106–116.

Following up on his preliminary report (see **#393**), Powell testifies before the Senate Committee on Appropriations regarding the Irrigation Survey's first year of operation. Powell presents a completed map of a section of the Rio Grande del Norte to demonstrate and explain the work of the survey.

390."The Lesson of Conemaugh." *North American Review* 149 (1889): 150–156.

In his response to the Johnstown Flood of 1889 that killed 2,000, Powell attempts to reassure the public that reservoirs should continue to be built. The work of the Irrigation Survey will ensure that dams are constructed adequately to meet the surrounding natural conditions. In *Seeing Things Whole* (**#213**).

391."Major Powell's Address." *Official Report of the Proceedings and Debates of the First Constitutional Convention of North Dakota, Assembled in the City of Bismarck, July 4–August 17, 1889.* Bismarck, ND: Tribune State Printers, 1889. 410–412.

Powell delivered this speech to the North Dakota Constitutional Convention on 5 Aug. 1889 while traveling through the West as a member of the Senate's Special Committee on the Irrigation and Reclamation of Arid Lands. Powell urges delegates to consider that arid lands depend on irrigation; therefore, they should fix in their constitution that corporations and individuals should not own property with exclusive water rights. Water should be held communally by the people of the state for the benefit of all. Reprinted in 1936 (**#867**) and 1969 (**#567**).

392."Ninth Annual Report of the Director of the United States Geological Survey." *Ninth Annual Report of the United States Geological Survey, 1887–'88.* Washington: GPO, 1889. iii–xiii, 1–46.

In addition to progress reports, Powell writes eulogies for F. V. Hayden, Roland Irving, James Stevenson, and Thomas Hampson.

Warman separate not seen.

393."Preliminary Report on the Organization and Prosecution of the Survey of the Arid Lands for Purposes of Irrigation." *Letter From the Secretary of the Interior, Transmitting, in Pursuance of Law, Report of the Geological Survey on the Subject of Irrigation.* 50th Cong., 2nd Sess., S. Ex. Doc. 43 Serial 2610. Washington: GPO, 1889. 3–12.

Powell's report, dated 31 Dec. 1888, is directed to the Senate Committee on Appropriations, which was considering refunding the Irrigation Survey after its first year of operation. Powell remarks on the importance of the Irrigation Surveys and presents an estimate of appropriations needed for the continuance of the work. He calls for the withdrawal of public lands in the arid regions, except those selected as irrigable lands. Reprinted in 10th annual report of the USGS (**#416**).

394."Report [on the Effect Upon Certain Rivers in Colorado of the Diversion of Water for Irrigation]." *Letter From the Secretary of the Interior, Transmitting, in Response to Senate Resolution of August 29, the Effect Upon Certain Rivers in Colorado of the Diversion of Water for Irrigation.* 50th Cong., 2nd Sess., S. Ex. Doc. 120 Serial 2612. Washington: GPO, 1889. 2–6.

In response to the Senate's request for information on the effect of diverting water from the Platte and Arkansas rivers on landowners further downstream, Powell states that the use of these rivers for agricultural purposes "will have no practical effect upon their uses as powers in Kansas and Nebraska."

1890

Articles

395."Bureau of Ethnology." *Annual Report of the Board of Regents of the Smithsonian Institution ... 1888.* Washington: GPO, 1890. 62–75.

Largely the same as the Director's Report in the BAE's annual report for the year (**#434**).

396."Bureau of Ethnology." *Annual Report of the Board of Regents of the Smithsonian Institution ... 1889.* Washington: GPO, 1890. 55–65.

Largely the same as the Director's Report in the BAE's annual report for the year (**#450**).

397."The Cañons of the Colorado." *The Normal Course in Reading: Fifth Reader.* New York: Silver, Burdett, 1890. 33–41.

A condensed version of the 1869 expedition for younger readers.

398."Dr. Cook as a Geologist." *Rutgers Targum* 10 (1890): 207–208.

Delivered at the memorial services of George Hammell Cook held 17 June 1890. Reprinted in 1891 (**#432**).

399."The Humanities." *Forum* 10 (1890): 410–422.

The process of human evolution is cultural, not biological. The study of mankind and his cultural activities is anthropology, or the science of the humanities. Powell provides a hierarchical system of classification for the "anthropologic" activities: technology, psychology, sociology, and philosophy. Through human endeavor, the law of "survival of the fittest" is replaced by laws of culture. This is not the same as article with the same title in *Science* (**#476**).

400."Institutions for the Arid Lands." *Century Illustrated Monthly Magazine* 40 (1890): 111–116.

Third of a series, with "The Irrigable Lands" (**#401**) and "The Non-Irrigable Lands" (**#406**). The essence of Powell's most radical ideas is found in this article. He argues that governance and ownership of arid lands should be organized by natural watersheds. Land ownership by neither small farmers nor wealthy capitalists can realize the potential for agricultural development of the irrigable lands. Irrigation districts should be formed, and all the people within the natural boundaries of hydrographic basins should make their own laws to divide and protect the waters, forests, and pastures. The federal government should keep its hands off governance and instead survey the land to establish and distribute the districts. "This, then, is the proposition I make: that the entire arid region be organized into natural hydrographic districts, each one to be a commonwealth within itself." The text of this article was delivered as an address to the New York Chamber of Commerce on 3 Dec. 1889 (quoted in *The New York Times* 4 Dec. 1889). In *Seeing Things Whole* (**#213**).

401."The Irrigable Lands of the Arid Region." *Century Illustrated Monthly Magazine* 39 (1890): 766–776.

First of a series, with "The Non-Irrigable Lands" (**#406**) and "Institutions for the Arid Lands" (**#400**). In these three articles, Powell takes the substance of his *Report on the Lands* and develops his ideas further for a general audience. He describes various "sub-humid" regions of the West, beginning at the 100th meridian, that he considers irrigable. Building his case, he provides numbers: amount of land needing irrigation, amount of water available, and cost per acre for "redemption of the arid region." The potential for the West is in agriculture, but there are major conflicts brewing over water rights with the current system of land ownership. Powell calls for four types of surveys—topographic, hydrographic, engineering, and geologic—so that the best sites might be chosen for maximum utilization of available water. See also **#402**.

402."The Irrigable Lands of the Arid Region." *Public Opinion* 8 (1890): 499–500.

Extract from **#401**.

403."Irrigation. Statement of Maj. J. W. Powell, Director of the Geological Survey." *Sundry Civil Appropriation Bill.* 51st Cong., 1st Sess., H. Rept. 2407 Serial 2813. Washington: GPO, 1890. 53–75.

Powell's testimony before the House Committee on Appropriations, dated 4 June 1890, is included in this report accompanying the appropriation Bill H.R. 10884. Powell once again defends the Irrigation Survey against Congressional attack. The testimony of Dutton regarding the necessity of topographical mapping is used against Powell by Senator Moody, who entangles himself in the deliberation of the House committee. See also the Senate version (**#415**).

404."Major Powell's Address to the Mining Engineers." *Science* 15 (1890): 156–157.

Paper delivered in an address to the American Institute of Mining Engineers in Washington. Powell hails the engineers as constituents among the body of "modern rulers who are the men who control the powers of nature." The USGS, in its effort to map the United States, works in cooperation with the mining engineers to advance scientific industry and extract riches from the earth.

405. "Major Powell's Reply." *New York Herald* 12 January 1890.

The "Cope-Marsh Controversy" stands as one of the most publicly embarrassing episodes of petty jealousy and backstabbing among prominent scientists. In this issue of the *Herald*, under the banner "Scientists Wage Bitter Warfare," Edward Cope publicly accuses O. C. Marsh of plagiarism and theft, Powell of incompetence, and the USGS of scientific monopoly. The *Herald* sent Powell the article ahead of time so that his reply could be published in the same issue. Powell's response, in the form of this letter, fields accusations with facts, counter-charges, and deadly sarcasm. As for Cope, "The paper to which I am now replying is sufficient evidence to any fair minded man that Professor Cope's mental and moral characteristics unfit him for any position of trust and responsibility."

406."The Non-Irrigable Lands of the Arid Region." *Century Illustrated Monthly Magazine* 39 (1890): 915–922.

Second of a series, with "The Irrigable Lands" (**#401**) and "Institutions for the Arid Lands" (**#400**). Turning his attention to the areas of the American West that cannot be redeemed for agriculture, Powell describes the potential of such land to support industries related to timber, pasturage, and mining. But the forests are in danger of disappearing, because man, in his effort to protect forests by quenching natural burning, has allowed inflammable materials in the form of brush, leaves, and other debris to accumulate and feed wildfires. Powell concludes that the industries of the West are unlike those of the East, thus new institutions are necessary. In *Seeing Things Whole* (**#213**).

407."Note by the Director." *A Dakota-English Dictionary*. By Stephen Return Riggs. Contributions to North American Ethnology, Vol. 7. Washington: GPO, 1890. v–viii.

Background regarding Riggs's work for this publication, which is an enlarged edition of a previously published dictionary (1852). Powell comments particularly on the illustrations and the organization of the report.

408."The Personal Characteristics of Professor Baird." *Smithsonian Annual Report for 1888*. Pt. I. Washington: GPO, 1890. 739–744.

See **#361** for annotation and references. Warman separate not seen.

409."Prehistoric Man in America." *Forum* 8 (1890): 489–503.

Citing evidence from archaeology, Powell sets out to prove that the American Indian is indigenous to North and South America. There is no scientific evidence that such physical remains as pueblos and mounds are the work of a highly "civilized" people.

410."Problems of American Archaeology." *Forum* 8 (1890): 638–652.

Powell describes several fields of research in which American archaeologists and anthropologists are engaged, and he decries the lack of trained workers in those fields. He calls for universities and scholars to turn their attention from the Orient to America, especially since the American Indians should be studied before they are completely assimilated. He points out the importance of classifying North American Indians according to language. See **#411**.

411."Problems of American Archaeology." *Public Opinion* 8 (1890): 429–430.

Extract from **#410**.

412."Remarks of Maj. John W. Powell, Director of the U.S. Geological Survey." *Three Americas Exposition: Arguments Before the Quadro-Centennial Committee of the United States Senate in Support of Senate Bills Nos. 1839 and 1135, Each Entitled "A Bill to Provide for a Three Americas and World's Exposition at the National Capital in 1892.*" Washington: GPO, 1890. 16–20.

Hearings held 10 Jan. 1890. Powell advocates federal support of the world exposition proposed for Washington. Exposition buildings should be built for permanent occupation by various government departments. Exhibits of mineral resources, natural history, and archaeology should be created to "have permanent value."

413.["Report Relative to the Use Made of the Appropriation for the Irrigation Survey"]. *Letter From the Secretary of the Interior Transmitting, in Response to Resolution of May 26, 1890, a Report Relative to the Use Made of the Appropriation for the Irrigation Survey.* 51st Cong., 1st Sess., S. Ex. Doc. 141 Serial 2688. Washington: GPO, 1890. 2–8.

Senator Stewart pushed through a resolution demanding to know whether irrigation appropriations had been misappropriated for topographic surveys. Powell's statement quotes from previous Congressional testimony, reports, and statutes to support his position that the operations of the USGS and the Irrigation Survey were distinct and that appropriations for the Irrigation Survey were used exclusively for that purpose.

414."Statement of J. W. Powell, Director of the Geological Survey [to the Special Committee on Irrigation and Reclamation of Arid Lands]." *Report of the Special Committee of the United States Senate on the Irrigation and Reclamation of Arid Lands.* Vol. 4. 51st Cong., 1st Sess., S. Rept. 928 Serial 2708, Pt. 5. Washington: GPO, 1890. 5–95, 151–204.

Powell's principal statements were made on 17 Jan. and 14 Mar. 1890. He came prepared with maps and data to present facts in relation to arid lands in order to present his plan for settlement based on irrigation districts. Each district would be self-governed. Using the Rio Grande River basin as an illustration of "the facts and conditions pertaining to every great river on the arid lands," Powell points out that various schemes have begun on the river valley to remove one half of the water supply. To avoid problems caused by such ill-conceived projects and to allow optimal agricultural development, the irrigable land should be defined and the people farming those lands should have the right to control the catchment areas and use the forests and grasses. Artesian wells are another source of water. But interstate and international boundary disputes and other pressures need to be resolved before agriculture can properly develop in the arid regions. The West is not like the East: "New conditions demand new institutions." The senators frequently interrupt Powell and a good deal of his statement is diverted by hostile questions from Senator Stewart regarding topographic mapping.

This volume of the *Report of the Special Committee ...* (**#835**) is subtitled "Statements by Director Powell and other officers of the U. S. Geological Survey. Consular reports. General report on irrigation in United States." Analogous House testimony is found in 11th annual report of the USGS (**#419**) and *Ceding the Arid Lands* (**#428**).

415."Statement of Major J. W. Powell, Director of the Geological Survey." *Sundry Civil Appropriation Bill.* 51st Cong., 1st Sess., S. Rept. 1466 Serial 2710. Washington: GPO, 1890. 46–109, 131–136.

Powell defends the Irrigation Survey from budget cuts before the Senate Committee on Appropriations. Powell concludes with an argument for disposition of public lands following completion of the survey. If the arid lands are reopened immediately, corporations will control the waters, thus inhibiting cultivation. If the surveying is completed before settlement and the land is parceled out in small parcels, cultivation will follow more quickly. See also the House version (**#403**).

416."Tenth Annual Report of the Director of the United States Geological Survey." *Tenth Annual Report of the United States Geological Survey, 1888–'89.* Washington: GPO, 1890. Pt. I, iii–xv, 1–80; Pt. II, i–viii, 1–65.

Part I reports on topographical and geological work of the USGS. Additional employees were added to absorb extra work associated with the Irrigation Survey. The geological work was reorganized according to geological provinces rather than geological groups. "Conference on map publication" (pp. 56–79) is a report on a conference held in January 1889 for federal geologists on a cartographic plan for publication of geologic maps and sections. Powell's report addresses in detail the nomenclature, conventional symbols for geologic maps, and conventional symbols for geologic sections.

Part II is the first annual report of the Irrigation Survey through June 1889. Resolutions bearing upon the initiation of the Irrigation Survey are included, along with previously published Congressional testimony (**#369**, **#393**). Powell presents his views on the disposal of public land, stating that water should be attached to the land and segregated with it. An irrigation survey is necessary to select and reserve sites for dams, canals and reservoirs. Non-irrigable land should be withdrawn from entry to forestall speculators. There is too little water available to permit irrigating all lands.

Warman separate not seen.

1891

Monographs

417.*Linguistic Stocks of American Indians North of Mexico.* Washington: BAE [New York: Sackett & Wilhelms Litho.], 1891.

This map, which accompanies the article "Indian Linguistic Families of America North of Mexico" in the 7th annual report of the BAE (#**422**), was issued separately as a limited edition for the use of students. It shows the habitat of the families at the time they were first known to Europeans. It was revised and published in the *Report on Indians Taxed and Not Taxed in the United States at the Eleventh Census, 1890.* The BAE again revised the map in 1906, in advance of its publication in *Handbook of American Indians North of Mexico* (#**812**) and yet again in 1915.

Articles

418."Eighth Annual Report [of the Director] of the Bureau of Ethnology." *Eighth Annual Report of the Bureau of Ethnology to the Secretary of the Smithsonian Institution 1886–'87.* Washington: GPO, 1891. xvii–xxxvi.

The usual report on fieldwork, office work, and publication. Warman separate not seen.

419."Eleventh Annual Report of the Director of the United States Geological Survey." *Eleventh Annual Report of the United States Geological Survey, 1889–'90.* Washington: GPO, 1891. Pt. I: iii–xv, 1–30; Pt. II: i–xiv, 1–289, 345–388.

In Part I, Powell reports on two administrative changes for the year. Because of the charge by Congress to conduct a hydrographic survey of the arid West, Powell delegated some administrative responsibility for the geological work of the USGS to G. K. Gilbert. Appropriations for engraving sheets of the atlas of the United States were transferred from the Public Printer and a Division of Engraving was organized within the USGS.

Part II is the second (and last) annual report of the Irrigation Survey. It consists of "Hydrography" and "Engineering," which describe the scope, methods, and activities of the survey; a catalogue of irrigation literature; and "The Arid Lands," which contains testimony presented by Powell before the House Select Committee on Irrigation in 1890 (for annotation, see **#428**). Also includes a report by A. H. Thompson on the topographic work of the Irrigation Survey

Warman separate not seen.

420."The Four Modes of Life." *Forum* 10 (1891): 667–677.

Continuing "The Growth of Sentiency" (**#421**), Powell describes and illustrates the transformations that occur through each of the four stages of evolution for each of the four modes of life, which are plant life, animal life, psychic life, and social life. But the four lines of evolution are not parallel. Instead, evolution itself evolves, and at its apex there develops a system of ethics, composed of laws of conduct, which obviates the law of "survival of the fittest."

421."The Growth of Sentiency." *Forum* 11 (1891): 157–167.

Nature gives more life than it can support. The law of evolution that governs biological life is competition, or "the survival of the fittest in the struggle for existence." The rate of evolution slows as beings become more complex and developed. Then a second law of evolution takes effect for sentient life, which is the struggle for happiness. There are four stages of evolution: vitality, sentiency, percipiency, and volitiency (see **#420**).

422."Indian Linguistic Families of America North of Mexico." *Seventh Annual Report of the Bureau of Ethnology to the Secretary of the Smithsonian Institution 1885–'86*. Washington: GPO, 1891. 1–142.

Powell's linguistic families formed the basis of future linguistic work among North American Indians. To this day, the main outline remains with few changes. This important work had its foundation in the earlier ethnological work of Albert Gallatin. Powell revises Gallatin, basing much of his revision on the work of ethnologists and linguistics associated with the BAE. "Rules of Nomenclature" details seven "rules" for naming the linguistic families. Powell treats the classification of languages in terms of lexical elements; that is, linguistic relationships are determined by similarities between vocal elements. "Literature Relating to the Classification of the Indian Languages" is an annotated bibliography of works dating back to 1836. "Linguistic Map" explains the accompanying map (*Linguistic Stocks of American Indians North of Mexico*; see **#417**). After describing the current Indian population and classes of tribal lands, Powell concludes that prior to the Europeans, tribes were mainly sedentary. Their languages belong to 58 distinct families with no apparent unity of origin. The families are enumerated in alphabetical order and accompanied by a table of synonyms of the family name, geographic area occupied, and principal tribes of each family. A summary of this paper was published in *Science* (**#429**). See reprints (**#577**, **#590**).

423. "The New Lake in the Colorado Desert." *Science* 18 (1891): 65–66.

Remarks reprinted from statements quoted in *The New York Times* (22 June 1891). There are many basins in the deserts of California that have been formed by portions of the Colorado River delta being cut off and left to evaporate. The recent overflow of the Colorado into the Coahuila Basin is not enough to fill the basin. See also **#424**.

424.["The New Lake in the Colorado Desert"]. *Journal of the American Geographical Society*. 23 (1891): 376–377.

Same as **#423**.

425."The New Lake in the Desert." *Scribner's Monthly* 10 (1891): 463–468.

Powell explains the formation and flooding of the Coahuila Basin in the desert of southern California. In so doing, he refutes the notion that the lake would give off evaporation, bringing humidity to the desert. However, in the future it may be possible to control and utilize the Colorado River above its delta, where it flows "between firm banks … by headworks of masonry, from which a long canal could lead to the agricultural land."

426."Report of the Director of the Bureau of Ethnology." *Annual Report of the Board of Regents of the Smithsonian Institution ... 1890*. Washington: GPO, 1891. 47–54.

Largely the same as the Director's Report in the BAE's annual report for the year (**#455**).

427."Seventh Annual Report [of the Director] of the Bureau of Ethnology." *Seventh Annual Report of the Bureau of Ethnology to the Secretary of the Smithsonian Institution 1885–'86*. Washington: GPO, 1891. xv–xli.

In "Explorations in Stone Villages" (xviii–xxiv), Powell reports on his fieldwork with James Stevenson in northern Arizona and New Mexico in 1885. See also "Indian Linguistic Families of America North of Mexico" (**#422**), an accompanying paper for this year's report.

Warman separate not seen.

428."Statement of Maj. J. W. Powell, Director of the Geological Survey." *Ceding the Arid Lands to the States and Territories.* 51st Cong., 2nd Sess., H. Rept. 3767 Serial 2888. Washington: GPO, 1891. Appendix B, 11–202.

Ceding the Arid Lands is the majority report of the House Select Committee on Irrigation of Arid Lands. Appendix B contains transcripts of testimony by Powell, Dutton, and others from 27 Feb. 1890 on a bill proposed by Hilary Herbert to transfer all public lands to the states, opening them once again to settlement and irrigation. In prepared statements and exchanges with committee members, Powell presents his vision for the West and a wealth of information about watersheds throughout the region. Published in the 11th annual report of the USGS as "The Arid Lands" (**#419**). This is analogous to Powell's testimony before the Senate (see **#414**).

Warman separate not seen.

429."The Study of Indian Languages." *Science* 17 (1891): 71–74.

Summary of "Indian Linguistic Families of America North of Mexico" (**#422**).

430."Twelfth Annual Report of the Director of the United States Geological Survey." *Twelfth Annual Report of the United States Geological Survey, 1890–'91.* Pt. I. Washington: GPO, 1891. iii, xii, 1–19.

Progress reports on the topographic, geologic, paleontologic, and accessory work of the USGS.

Warman separate not seen.

Abstracts, Remarks

431. Address. *American Geologist* 8 (1891): 256–257.

Abstract of a presentation on map coloring and cartography at the Washington meeting of the International Congress of Geologists in August 1891.

Reprints

432."Dr. Cook as a Geologist." *Addresses Commemorative of George Hammell Cook Etc.* Newark, NJ: Advertiser Printing House, 1891. 20–27.

Reprint of **#398**. Not seen.

1892

Articles

433."National Agencies for Scientific Research." *Chautauquan* 14 (1892): 37–42, 160–165, 291–297, 422–425, 545–549, 668–673.

> Each section of this serial article discusses the organization, work, and achievements of a different government agency: Smithsonian Institution, Dept. of Agriculture, Weather Bureau, Coast and Geodetic Survey, and ending with the USGS. The work of these agencies demonstrates that science contributes to the welfare of the United States. As befitting a highly civilized nation, such agencies should be supported by the government.

434."Ninth Annual Report [of the Director] of the Bureau of Ethnology." *Ninth Annual Report of the Bureau of Ethnology to the Secretary of the Smithsonian Institution 1887–'88*. Washington: GPO, 1892. xxiii–xlvi.

> Powell mentions his archaeological fieldwork in the Tewan Mountains. Warman separate not seen

435."The Nomenclature and Teaching of Anthropology [Remarks by J. W. Powell]." *American Anthropologist* 5 (1892): 266–271.

> Powell comments on a paper read by Daniel Brinton, disagreeing with Brinton's premise that scientific bodies should legislate nomenclature and classification for anthropology. Powell then proposes that his own classification scheme and nomenclature be adopted. Reprinted in De Laguna (**#801**).

436."Our Recent Floods." *North American Review* 155 (1892): 149–159.

> Describes some of the causes and types of flooding, methods for prevention, and precautions people can take. Powell makes the case for accurate topographic maps, geological surveys of floodplains, and the need for hydraulic surveys. He calls for expanding the Weather Bureau, particularly in forecasting flood and storm warning services.

437."Statement of Mr. John W. Powell, Director of the U.S. Geological Survey." *Report: [to Accompany S. 1549 Providing for the Public Printing and Binding and the Distribution of Public Documents.]* 52nd Cong., 1st Sess., S. Rept. 18 Serial 2911. Washington: GPO, 1892. 220–234.

> Along with other federal officials, Powell defends his department's printing budget. In usual fashion, Powell supplies the Senate

Committee on Printing with a detailed account of USGS printing costs and controlled distribution of publications, along with a well-reasoned justification for his publication program. He notes, "There is a large body of people in the country who read the reports of the Survey and who use them in the practical industries of the country."

438."Thirteenth Annual Report of the Director of the United States Geological Survey." *Thirteenth Annual Report of the United States Geological Survey, 1891–'92.* Pt. I. Washington: GPO, 1892. iii–vii, 1–66.

Powell presents a detailed plan of operation for the year with corresponding budget allocations and salaries. In addition, he acknowledges the receipt of the Cuvier Prize from the Academy of Sciences of the Institute of France in "Award of the Cuvier Prize by the Academy of Sciences of the Institute of France" (pp. 62–66). The prize was awarded in 1891 for the "collective work of the Geological Survey." Powell used the prize money to commission a gold model as a "souvenir" to be retained by the USGS. The remaining balance was used to establish a "Cuvier fund."

Warman separate not seen.

1893

Articles

439. Address and Remarks. *Official Report of the International Irrigation Congress Held at Los Angeles, California, October, 1893.* Los Angeles: Los Angeles Chamber of Commerce, 1893. 107–116.

Powell was the featured speaker at the Congress on 13 Oct. 1893, but he did not deliver his prepared technical paper on irrigation, later published as "The Water Supplies in the Arid Region" (**#464**). Instead, he argues that "not one more acre of land should be granted to individuals for irrigation purposes; there is not water enough." Water should be reserved for irrigable land already in the hands of individuals. Powell's response to intense questioning by angry delegates follows the text of his address.

440."Are Our Indians Becoming Extinct?" *Forum* 15 (1893): 343–354.

Before the white man arrived, Powell theorizes, the American Indians were widely scattered throughout the continent. They lived relatively peaceful lives and developed diverse cultures. Their population is now diminished, but many have made significant progress in moving out of savagery to becoming absorbed into modern civilization. The "conquering race" should continue its efforts to educate and civilize the remaining "lowest tribes."

441."Are There Evidences of Man in the Glacial Gravels?" *Popular Science Monthly* 43 (1893): 316–326.

In response to several previous articles in *Popular Science* regarding a dispute over the origin of man in North America, Powell explains how geologists employed by the USGS determine the age of gravel deposits. Numerous examples are given of alleged archaeological evidence of man's Ice Age habitation of North America that were subsequently dismissed by geological evidence. Scientific data cannot be ignored. See also editorial comments by Dana (**#825**).

442."Fourteenth Annual Report of the Director of the United States Geological Survey." *Fourteenth Annual Report of the United States Geological Survey, 1892–'93.* Pt. I. Washington: GPO, 1893. 3–165.

Following severe budget cuts by Congress, Powell's plan for the year notes that salaries for several division chiefs lapsed and they voluntarily continued their work without pay. The report includes a memorial to John Strong Newberry, "General Work in Taxonomy" (**#443**), "General Summary of Paleontologic Work [1879–1892]," and "General Summary of Work in Terrestrial Physics, 1880–'92."

Warman separate not seen.

443."General Work in Taxonomy." *Fourteenth Annual Report of the United States Geological Survey, 1892–'93.* Pt. I. Washington: GPO, 1893. 65–122.

In this essay embedded in the Director's Report (**#442**), Powell outlines his philosophy of the relationship of research to scientific knowledge. Knowledge, science, and classification progress in stages. The first stage is observation or experimental, in which facts are gathered. As accumulated facts or conceptions increase, eventually they form an "intellectual magma representing a certain plane of knowledge." It has been the policy of the USGS to employ research in order to "elevate the science of the earth and its resources to a higher plane."

444."The Geologic Map of the United States." *Transactions of the American Institute of Mining Engineers* 21 (1893): 877–887.

Read at the Montreal meeting of the Institute by C. W. Hayes. Powell describes the process of preparing a geologic map of the United States. A classification system for rocks and formations has been created and most of the country is classified. The USGS has developed a system of geological conventions. To date, one fifth of the nation has been surveyed and 700 atlas sheets completed.

445."History of Irrigation." *Independent* 45 (1893): 594–595.

This is the first of 11 essays in a series of articles in this issue of *The Independent*. Powell states that the earliest agriculture began in arid regions on land that could be irrigated by nearby smaller streams. In the United States, Mormons learned of irrigation practiced by Mexicans and Indians, leading to the reclamation of desert Utah. About two-thirds of the continental United States requires artificial irrigation to grow crops. Future development of the nation depends on three sources of little used water supply: the great rivers, reservoirs, and artesian waters.

446."The Mineral Exhibits at Chicago." *British Trade Journal* 31 (1893): 520–522.

Unsigned. A descriptive piece highlighting several exhibits at the Columbian Exposition in Chicago in 1893, including the USGS exhibit on mineral wealth in the United States. See reprint **#453**.

447."Report of the Director of the Bureau of Ethnology for the Year Ending June 30, 1891." *Annual Report of the Board of Regents of the Smithsonian Institution ... 1891*. Washington: GPO, 1893. Appendix I, 29–37.

Largely the same as the Director's Report in the BAE's annual report for the year (**#462**).

448."Report of the Director of the Bureau of Ethnology for the Year Ending June 30, 1892." *Annual Report of the Board of Regents of the Smithsonian Institution ... 1892*. Washington: GPO, 1893. Appendix I, 49–58.

Largely the same as the Director's Report in the BAE's annual report for the year (**#503**).

449."Simplified Spelling." *American Anthropologist* 6 (1893): 193–195.

One of 12 addresses delivered at a symposium titled "Is Simplified Spelling Feasible." The symposium took up three consecutive meetings of the ASW in 1892 and involved professors from nearby colleges, the Library of Congress, and others. Powell advocates a partial reform of the spelling of the English language to allow more time in the classroom to be spent on other subjects than spelling yet allow books in print to not become obsolete.

Warman separate not seen.

450."Tenth Annual Report [of the Director] of the Bureau of Ethnology." *Tenth Annual Report of the Bureau of Ethnology to the Secretary of the Smithsonian Institution 1888–'89*. Washington: GPO, 1893. ix–xxx.

Highlights include a eulogy for James Stevenson (xxiv–xxv) and comments on the accompanying paper by Garrick Mallery, "Picture-Writing of the American Indians," which updates Mallery's preliminary report on the topic presented in the 4th annual report of the BAE. This "work is unique because it presents the several anthropologic topics recorded by the Indians themselves according to their unbiased conceptions, and in their own mode of writing." Powell's comments were used as the foreword to a 1972 reprint of "Picture-Writing" (**#579**).

Warman separate not seen.

451."The Work of the U.S. Geological Survey." *Science* 21 (1893): 15–17.

Read before the Geological Society of America at Ottawa, 30 Dec. 1892. Powell gives the origin and organization of the USGS, then describes its principal current work: the geologic survey, topographic survey, and the collection and publication of information concerning mineral resources and mining. Future plans include a topographic map of the entire country and a geological atlas of the United States. According to Mary Rabbitt, the paper was read by WJ McGee, but "was not written by Powell or by McGee" (**#859**, p. 217).

Remarks

452."Discussion sur le Coloriage des Cartes Géologiques." *Congrès Géologique Internationale, Compte Rendu de la 5me Session, Washington, 1891.* Washington: GPO, 1893. 79–80, 208–211.

Powell opens a session of the Congress by exhibiting charts displaying the color schemes recently adopted for the general geological maps published by the USGS.

Reprints

453."The Mineral Exhibits at Chicago." *Kuhlow's German Trade Review, and Exporter (Berlin)* 18 (1893): 4503–4504.

See **#446**. Not seen.

1894

Articles

454. Address. *Proceedings of the New York Farmers, Season 1893–1894*. New York: Globe Stationery and Printing, 1894. 70–79.

Powell applies his knowledge of irrigation to its potential for eastern farming. Looking to the future, Powell predicts intensive agriculture on smaller parcels of land in suburban communities, using glass (greenhouses) and irrigation even in humid areas. Modern inventions are changing industry and rapidly transforming society.

Warman separate not seen.

455."Eleventh Annual Report [of the Director] of the Bureau of Ethnology." *Eleventh Annual Report of the Bureau of Ethnology to the Secretary of the Smithsonian Institution, 1889–'90*. Washington: GPO, 1894. xxi–xlviii.

Powell reports that some lines of research have been suspended in order to complete other investigations.

Warman separate not seen.

456."Immortality." *Open Court* 8 (1894): 4335–4337.

Part song, part blank verse, this poem sings the praises of man's successful endeavor to free himself from the bonds of primitive existence.

Warman separate not seen.

457."The Irrigation Convention." *Harper's Weekly* 38 (1894): 268.

A synopsis of the issues likely to come under discussion at the upcoming interstate irrigation convention in Omaha: the climate, problems of hydraulic engineering, and who should control the lands that might serve as catchment areas for irrigation districts. The article is accompanied by photographs of irrigation sites.

458."The North American Indians." *The United States of America: A Study of the American Commonwealth*. Vol. 1. Ed. Nathaniel Shaler. New York: Appleton, 1894. 190–272.

This lengthy article summarizes Powell's knowledge and theories regarding the Native Americans. Topics are the origin of the Indians, social organization, mythology, population distribution, language, music, and more. Includes a reproduction of the map *Linguistic Stocks of American Indians North of Mexico* (**#417**).

459."On the Nature of Motion." *Monist* 5 (1894): 55–64.

Powell assumes the theory that conservation of energy is a unifying principle within the law of physics. Using a rolling ball to demonstrate the various types of motion and the physical forces that act upon an object, Powell attempts to prove that a universal law of physics is "acceleration is deflection." Motion is inherent in matter.

460."Ownership of Lands in the Arid Region." *Irrigation Age* 6 (1894): 143–149.

Building on scientific data presented in the maps and charts of the first article in the series, "The Water Supplies in the Arid Region" (**#464**), Powell here provides comprehensive statistics on land ownership and usage for the western United States. Evidence points to inadequate water supplies available to irrigate privately owned lands in most states. The American people must decide whether available water should be held for use by landowners or diverted to public lands.

461."Report of the Director of the Bureau of Ethnology for the Year Ending June 30, 1893." *Annual Report of the Board of Regents of the Smithsonian Institution ... 1893*. Washington: GPO, 1894. Appendix 2, 38–44.

Largely the same as the Director's Report in the BAE's annual report for the year (**#495**).

462."Twelfth Annual Report [of the Director] of the Bureau of Ethnology." *Twelfth Annual Report of the Bureau of Ethnology to the Secretary of the Smithsonian Institution 1890–'91*. Washington: GPO, 1894. xxi–xlviii.

In the section titled "Accompanying Paper on the Mound Explorations of the Bureau" (xxxix–xlviii), Powell remarks on the change in direction of the BAE's work necessitated by a Congressional appropriation to investigate Indian mounds in the East. He recalls his own investigations of prehistoric mounds in Ohio, Indiana, Illinois, and Missouri in 1858–1860, continued during Civil War while campaigning in Missouri, Tennessee, and Mississippi. Questions raised during those years "seem to have reached a solution." Evidence shows that the prehistoric mounds were constructed by historical Indians rather than mythical vanished races.

463."The Water and the Method." *Proceedings of the Kansas Irrigation Association, at the Wichita Convention, November 22 and 23, 1893*. N.p.: n.p., [1894?]. 7–8.

This is a summation of Powell's address to the association. Using rainfall statistics for Kansas, Powell presents his argument that with proper surveying and planning, "the highest agriculture of America will ultimately be upon the arid and semi-arid lands of the 'Great American Desert.'" A footnote refers to "article illustrated and elaborated in *Irrigation Age*" (see **#464**).

464."The Water Supplies in the Arid Region; First Paper: 'The Duty of Water and the Sources of Supply.'" *Irrigation Age* 6 (1894): 54–65.

This article and "Ownership of Lands in the Arid Regions" (**#460**) were written at editor William Smythe's invitation, "as the outgrowth of the discussion aroused by Major Powell's celebrated speech at Los Angeles last October" (see **#439**). Using scientific data and maps, Powell describes the amount of water necessary to grow an acre of crop for one year (the "duty of water"). In discussing the potential sources of water supply, he introduces the terms "run-off" for water gathered into streams and "fly-off" for water lost to evaporation.

465."The Water Supply of the Great Plains and Its Availability for Irrigation Purposes." *Western America* 1.2 (1894): 6–9.

An address delivered 21 Mar. 1894 at the 2nd Annual Convention of the Inter-State Irrigation Association at Omaha, Nebraska. The official stenographer transcribed Powell's words literally, providing the reader with a good sense of Powell's speech cadence. Powell first discusses evaporation, rainfall, and the amount of rainfall available for irrigation. There must be catchment areas large enough to hold the amount of water necessary for successful irrigation. Other sources of water are artesian wells and reservoirs for storm waters. Catchment basins should be governed by the farmers who depend on them for irrigation. Harkening to his youth in Illinois, Powell calls for the establishment of small family farms: "I think that your happiness and our liberties depend more upon the establishment and preservation of these little homes than they do upon the great problems of business."

1895

Monographs

466.*Canyons of the Colorado*. Meadville, PA: Flood & Vincent, The Chautauqua Century Pr, 1895.

In the preface, Powell gives a history of this publication, which is based on the 1875 *Exploration*. He describes *Canyons* as a popular account of the 1869 expedition, revised and enlarged from his original journals and with six new chapters descriptive of the region and its people. He thanks all members of the first Colorado River expedition, including the Howland brothers and Dunn. There are over 250 illustrations from nearly every Powell Survey, including adaptations of sketches by William H. Holmes featured in the Dutton atlas (**#118**). The six new chapters include an overview of the physiography and ethnology of the Colorado Basin and an evocative concluding essay that expresses Powell's personal sentiment for the landscape.

See the introductory note for *Exploration* (**#242**); see also **#572**, **#575**, **#585**, **#588**, and **#592**.

467.*Physiographic Features*. National Geographic Monographs 1:2. New York: American Bk Co, 1895.

Along with other distinguished scientists, including G. K. Gilbert and Nathaniel Shaler, Powell authored three essays (dated March, April, and May 1895) in this 10-part series published by National Geographic to supplement geography textbooks. Powell's essays encompass the geography of the continental United States and are supplemented with illustrations by Moran, Hillers, and others. The 10 monographs were published in a single volume, *The Physiography of the United States* (**#504**). This second essay in the series addresses land and water forms, such as mountains and lakes; the processes by which they are produced; and the effect of fire, earthquake, and flood. See also **#468** and **#469**.

468.*Physiographic Processes*. National Geographic Monographs 1:1. New York: American Bk Co, 1895.

In this first of the three essays for National Geographic, Powell outlines the processes of vulcanism, diastrophism, and gradation. He introduces "three great moving envelopes" and their motion: atmosphere, water, and rock. See also **#467**, **#469**, and **#505**.

469.*Physiographic Regions of the United States.* National Geographic Monographs 1:3. New York: American Bk Co, 1895.

This essay has more in common with Powell's earlier writing on Western lands. Breaking with the past custom of describing physiographic regions in units of basins, Powell divides the nation into regions based on common physiographic characteristics, such as mountains, plateaus, and plains. Three of these regions are discussed in detail to illustrate the types: the Atlantic Plains, the Colorado Plateaus, and the Pacific Mountains. Powell's divisions were subsequently adopted by later geographers. See also **#467**, **#468**, and **#506**.

470.*Stone Art in America.* Washington: n.p., 1895.

By studying the geological evidence where stone implements have been found, American archeologists and geologists have concluded that the distinction between Paleolithic and Neolithic man "is not valid for this continent." This is a pamphlet edition of the article in *American Anthropologist* (**#485**). Reprinted in **#801**.

Articles

471."Anthropology." *Johnson's Universal Cyclopaedia.* New ed. Vol. 1. New York: A. J. Johnson, 1895. 233–234.

The science of man is divided into somatology, psychology, and ethnology. Various theoretical schools of anthropology and views on migration and racial classification are briefly identified.

472."Ethnology." *Johnson's Universal Cyclopaedia.* New ed. Vol. 3. New York: A. J. Johnson, 1894. 205–207.

Several definitions of ethnology are offered. Powell favors describing ethnology as the science that "treats of the culture of the tribes and nations of mankind." A series of sciences constitutes subdivisions within ethnology, such as technology, sociology, and philosophy. Many other terms are defined, consistent with Powell's idiosyncratic nomenclature.

473."Explanation." *Geologic Atlas of the United States.* USGS. Washington: GPO, 1895.

Powell's statement appears in the *Geologic Atlas*, folios 1–20, pages 2–3 of each folio published 1894–1895. Not seen.

474."Fifteenth Annual Report of the Director [of the United Sates Geological Survey]." *Fifteenth Annual Report of the United States Geological Survey, 1893–'94*. Washington: GPO, 1895. 9–108.

Powell announces his retirement from the USGS with a few words of farewell and special mention of his colleagues Stevenson, Pilling, Thompson, Gilbert, Walcott, and Hillers. In "The "Geologic Atlas" (pp. 79–90), he describes the plan, method, and progress of publication of the folios of the atlas.

Warman separate not seen.

475."The Five Books of History." *Science* ns 1 (1895): 157–161.

The endeavors of the past may be found in the Stone Book, the Ruin Book, the Tomb Book, the Folk Book, and the Scripture Book By looking to the five books for keys to our heritage, scientists will discover the materials from the past by which mankind continues to develop industries, pleasures, languages, institutions, and opinions. "The past is not a pall on the present, hiding the truth, but a searchlight that may be turned on the future." See also **#533**.

476."The Humanities." *Science* ns 1 (1895): 15–18.

The humanities are industries, pleasures, languages, institutions, and opinions. Mankind has evolved on these "five parallel roads ... toward the goal of wisdom." Animal evolution multiplies, but human evolution tends toward unification as human activities become more interrelated. The "old science of ethnology" should be limited to the study of "biotic" characteristics and the term "demology" should be used for the study of cultural activities.

477."Indians of North America." *Johnson's Universal Cyclopaedia*. New ed. Vol. 4. New York: A. J. Johnson, 1895. 544–552.

A comprehensive overview of religion and customs, arts and industries, languages, dwellings, and all aspects of North American Indians. The article concludes with a bibliography of published materials on the subject of North American Indian languages.

478. Letter. *In Memoriam: James Constantine Pilling, 1846–'95*. Washington: D. W. Gregory, 1895. 12–13.

A tribute to Pilling, written to his widow and reproduced in this privately printed pamphlet.

479."Mindellef, Cosmos." *Johnson's Universal Cyclopaedia*. New ed. Vol. 5. New York: A. J. Johnson, 1895. 778.

Brief biography of a BAE archaeologist.

480."Proper Training and the Future of the Indians." *Forum* 18 (1895): 622–629.

The U.S. government long ago embarked on a policy of managing the Indians by treating them as sovereign nations. While expensive, the policy has been advantageous to the Indians and should be continued. The government has an obligation to appropriate funds to establish industrial schools for Indian youth.

481. ["Report on the Methods of Surveying the Public Domain"]. *To Improve the Public Surveys; Report [From] the Joint Commission of Congress to Inquire Into Status of Laws Organizing the Executive Departments.* 53rd Cong., 3rd Sess., S. Rept. 1021 Serial 3148. Washington: GPO, 1895. 21–24.

Excerpt from Powell's 1878 report to the NAS on the topic of consolidating the surveys of the public lands (see **#261**). House version published as 53rd Cong., 3rd Sess., H. Rept. 1954 Serial 3158.

482."Scenery on the Colorado." *American Antiquarian* 17 (1895): 240–247.

Excerpt from *The Canyons of the Colorado*. Main points of interest and the geology and climate of the Colorado River region are described.

483."Shoshonean Indians." *Johnson's Universal Cyclopaedia*. New ed. Vol. 7. New York: A. J. Johnson, 1894. 507–509.

List of various tribes considered Shoshonean by linguistic stock. Includes bibliography ("Authorities").

484."The Soul." *Monist* 5. (1895): Appendix, 1–16.

An ode to the power of man's mind and will to use the universal forces of the soul—sense, perception, wisdom and reflection—to transform the world.

485."Stone Art in America." *American Anthropologist* 8 (1895): 1–7.

See **#470** for annotation. Reprinted in **#801**.

1896

Monographs

486.*On Primitive Institutions*. Philadelphia: Dando Printing & Publishing, 1896.

Pamphlet edition of an address read at the American Bar Association, meeting in Saratoga Springs, New York, on 21 Aug. 1896. This essay encompasses Powell's theory of human social organization. The character of primitive institutions may be seen in the manner in which regimentation and justice are realized. Regimentation, of which there are several types, is the grouping of peoples by institutional bonds; justice is the social function that such a grouping performs. The two major types of regimentation are tribal (primitive) and national (civilized). In the transition to national society, regimentation takes the form of justice.

Also published in ABA proceedings (see **#497**).

487.*Relation of Primitive Peoples to Environment, Illustrated by American Examples*. Washington: GPO, 1896.

One in a series of Saturday lectures in the USNM, delivered 25 Apr. 1896. In his later years, Powell's thoughts increasingly took a philosophical bent, resulting in articles like this. Veering from a discussion of human evolution to a consideration of the importance of language to social environment, Powell attempts to distinguish between quality and property. Using examples from everyday life, Powell demonstrates that properties are inherent in things themselves and cannot be changed, while qualities depend on a point of view. Offprint of the Smithsonian report for 1895 (**#499**).

488.*Seven Venerable Ghosts*. Washington: n.p., 1896.

Address of the retiring president of ASW, delivered 4 Feb. 1896. The "ghosts" are substrate, essence, space, force, time, spirit, and cause. Each of these has two meanings, one within the proper context of reality (science), the other within the improper context of metaphysics (illusion). The seven ghosts of science "have stalked through the history of philosophy … [and] have terrified mankind as potent realities." The philosophy of science builds on the known, while metaphysics or transcendental philosophy builds on the unknown. In distinguishing certitude from illusion, it is also important to distinguish between properties, which are real and grounded in reality, and qualities, which are subjective. Properties are constant and factual while qualities are relative and changeable.

Published in the *American Anthropologist* (**#502**). See also **#489** and **#510**.

Articles

Certitudes and Illusions: The following articles and correspondence from *Science* (**#489–#491**) are presented as a group in chronological order. In this series of articles, as in most of his remaining work, Powell attempts to construct a philosophy of science and explain the law of mental evolution. Without science and scientific thinking, Powell believes, mankind's progress is cut short and we are doomed to living under a cloud of illusion. Many of these articles appear in *Truth and Error* (**#510**).

489."Certitudes and Illusions: Chuar's Illusion." *Science* ns 3 (1896): 263–271.

Powell begins the first article with a story from his own experience in 1880, in which an Indian chief, Chuar, presents his subjective interpretation of gravity. Chuar's thinking represents illusion. Powell then attempts to define and categorize what is known about matter. Everything that can be known in the real world consists of substrate, essence, space, force, time, ghost, and cause. Each is expressed in terms of number, extension, motion, duration, and judgment of bodies in their relations. Things that cannot be thus expressed are illusions. For an example of responses to this article, see **#874**. See also **#488**, **#510**, and **#511**.

490."Certitudes and Illusions: An Illusion Concerning Rest." *Science* ns 3 (1896): 426–433.

Powell demonstrates that a body in motion may deflect its particles in a collision with another body, but the motion itself cannot be created or destroyed. A body that appears to be a rest is not motion destroyed but a body deflected.

491."Certitudes and Illusions [Discussion and Correspondence]." *Science* ns 3 (1896): 444–445.

Letter in response to criticism from a reader that "an anthropologist should express opinions concerning metaphysic." He points out that anthropology includes the study of the metaphysical and is subject to its own illusions.

492."Certitudes and Illusions [Discussion and Correspondence]." *Science* ns 3 (1896): 595–596.

Powell responds to questions raised in a letter to *Science*. He restates and clarifies statements in "Certitudes and Illusions: An Illusion Concerning Rest" (**#490**).

493."The Absolute and the Relative." *Science* ns 3 (1896): 743–745.

Another reply to correspondence regarding "Certitudes and Illusions: An Illusion Concerning Rest" (**#490**). Powell asserts that motion is both absolute and relative. In the physical world, there is neither pure motion nor pure rest.

494."The Subject of Consciousness." *Science* ns 3 (1896): 845–847.

In this response to a correspondent's remarks on the subject of consciousness, Powell includes an excerpt from a paper presented at a meeting of the NAS in which he states that a judgment is composed of consciousness and inference. Verification added by experience produces certitude.

495."Fourteenth Annual Report [of the Director] of the Bureau of Ethnology." *Fourteenth Annual Report of the Bureau of Ethnology to the Secretary of the Smithsonian Institution 1892–'93*. Pt. I. Washington: GPO, 1896. xxvii–lxi.

Powell reorganizes this and subsequent annual reports along the lines of his classification system for the philosophy of science: sociology, linguistics, mythology, and psychology. In his comments on the accompanying papers, he discusses the difference between scientific hypothesis, which is based on experience, and primitive hypothesis, which "rests on appeal to the unknown." The report includes a history of the Coronado Expedition (liv–lviii) and extensive comments on Powell's system of classification of Indian tribes, based not on somatology (biology) but on "purely human or demotic characters" (xxvii–xxx).

Warman separate not seen.

496."James Dwight Dana." *Science* ns 3 (1896): 181–185.

A memorial address given 14 Jan. 1896. Powell ends his eulogy with three stanzas from "Immortality" (**#456**).

497."On Primitive Institutions." *Report of the Nineteenth Annual Meeting of the American Bar Association Held at Saratoga Springs, N.Y., August 19, 20, and 21, 1896*. Philadelphia: Dando Printing, 1896. 573–593.

See **#486** for annotation.

498."Relation of Primitive Peoples to Environment, Illustrated by American Examples." *Relations of Human Life to Environment*. Smithsonian pamphlet No. 1064. Washington: GPO, 1896. 625–637.

See **#487** for annotation. This pamphlet is an offprint of pages 625–711 of the Smithsonian annual report for 1895 (**#499**).

499."Relation of Primitive Peoples to Environment, Illustrated by American Examples." *Annual Report of the Board of Regents of the Smithsonian Institution ... 1895*. Washington: GPO, 1896. 625–637.

See **#487** for annotation. See also **#498**.

500."Report of the Director of the Bureau of American Ethnology for the Year Ending June 30, 1894." *Annual Report of the Board of Regents of the Smithsonian Institution ... 1894*. Washington: GPO, 1896. Appendix 2, 44–57.

Largely the same as the Director's Report in the BAE's annual report for the year (see **#507**).

501."Report of the Director of the Bureau of American Ethnology for the Year Ending June 30, 1895." *Annual Report of the Board of Regents of the Smithsonian Institution ... 1895*. 1896. Appendix 2, 42–54.

Largely the same as the Director's Report in the BAE's annual report for the year (see **#509**).

502."Seven Venerable Ghosts." *American Anthropologist* 9 (1896): 67–91.

See **#488** for annotation.

503."Thirteenth Annual Report [of the Director] of the Bureau of Ethnology." *Thirteenth Annual Report of the Bureau of Ethnology to the Secretary of the Smithsonian Institution 1891–'92*. Washington: GPO, 1896. xxi–lix.

Powell mentions his visit to northern California to study Indian languages and reiterates some of his views on primitive philosophy.

Warman separate not seen.

Reprints

504."Physiographic Features." *The Physiography of the United States.* New York: American Bk Co, 1896. 33–64.

See **#467**.

505."Physiographic Processes." *The Physiography of the United States.* New York: American Bk Co, 1896. 1–32.

See **#468**.

506."Physiographic Regions of the United States." *The Physiography of the United States.* New York: American Bk Co, 1896. 65–100.

See **#469**.

1897

Articles

507."Fifteenth Annual Report [of the Director] of the Bureau of Ethnology. *Fifteenth Annual Report of the Bureau of Ethnology to the Secretary of the Smithsonian Institution 1893–'94.* Washington: GPO, 1897. xvii–cxxi.

In his introduction, Powell posits that primitive people should be studied as "superorganic groups, in which each individual reflects and is molded by the characteristics of his associates." The science that undertakes such studies is demonomy, the science of humanity. Includes the essay "On regimentation" (**#508**).

Warman separate not seen.

508."On Regimentation." *Fifteenth Annual Report of the Bureau of Ethnology to the Secretary of the Smithsonian Institution 1893–'94.* Washington: GPO, 1897. civ–cxxi.

Regimentation is the grouping of people by institutional bonds. There are two methods of regimentation: the tribal system, organized on the basis of kinship, and the national system, organized on the basis of territory. Feudal society is transitional between the two. Powell defines the various units within tribal society and their functions. In "Report of the Director" (**#507**).

509."Sixteenth Annual Report [of the Director] of the Bureau of American Ethnology." *Sixteenth Annual Report of the Bureau of American Ethnology to* *the Secretary of the Smithsonian Institution 1894–'95*. Washington: GPO, 1897. xv–xcix.

Continuing his discussion of demonomy as presented in the 15th annual BAE report (**#507**), Powell characterizes the five categories of human activities: esthetology, technology, sociology, philology, and sophiology. These categories can be used "for a scientific classification of the races and peoples of the earth." Powell's essays on these categories appear in the 19th and 20th annual reports of the BAE (see **#537**, **#557–#560**). Powell also writes obituaries for Garrick Mallery and James Owen Dorsey in "Necrology" (lxxx–lxxxv).

Warman separate not seen.

1898

Monographs

510. *Truth and Error, or The Science of Intellection*. Chicago: Open Court, 1898.

According to William M. Davis, Powell "planned to arrange the entire content of knowledge in a system of three parts, the first to deal with Nature, the second with Man, and the third with Mind ... left unfinished" (see **#611**, p. 81). *Truth and Error* is the first of the projected trilogy. The second, "Good and Evil," appeared as a series of articles in the *American Anthropologist*, 1899–1901 (see **#528**, **#531**, **#532**, **#541**, **#550**).

Underlying Powell's attempt to categorize all phenomena according to a set of essential properties is a classification system based on the number five. By discovering and utilizing the unifying principles of all existence, Powell's rational "Philosophy of Science" will replace metaphysics and other lesser philosophies. The book is dedicated to Lester Ward, who wrote a critical review (see **#879**).

Portions of *Truth and Error* appear in *Open Court* articles (see **#511**, **#513**, **#517**).

Articles

511. "Chuar's Illusion." *Open Court* 12 (1898): 577–581.

See also **#510**. Same as large portions of "Certitudes & Illusions: Chuar's Illusion" (**#489**).

512."The Evolution of Religion." *Monist* 8 (1898): 183–204.

Powell adds a new science to his study of the culture of mankind: "Demonomy, or the science of the Humanities." This essay serves well as a summary of Powell's publications on the evolution of the five humanities, though its primary purpose is to trace the evolution of religion through the four stages of society—savagery, barbarism, monarchy and democracy. As religion approaches perfection, creed and ceremony remain, but there is more scientific truth. Ultimately, ethical religion will triumph.

513."Fallacies of Perception." *Open Court* 12 (1898): 720–729.

Powell gives examples of optical illusions and discusses other types of illusions. Perception is the interpretation of a symbol. For example, a common erroneous perception is that a rainbow has a surface because humans, from habit and experience, interpret color as a mark of surface. Rather than dismiss such illusions, even when armed with facts, the mind will search for some metaphysical explanation. In one of his last direct stabs at metaphysics, Powell states that metaphysics patronizes science in its efforts to establish the existence of ghosts and other illusions but denigrates reality as "base-born."

Excerpted from *Truth and Error* (**#510**).

514."Forest Dwellers—Indians." *Nature and Art* 1 (1898): 48–51.

Description of sylvan tribes written for a juvenile audience.

515."How a Savage Tribe Is Governed." *Forum* 25 (1898): 712–722.

Powell describes the governance of tribal society, which consists of clans determined by rules of kinship. Marriage laws and other customs are enforced by councils of elders. Shamans manage the ceremonies relating to war and hunting and gathering.

516."An Hypothesis to Account for the Movement in the Crust of the Earth." *Journal of Geology* 6 (1898): 1–9.

Read at the November 1897 meeting of the NAS in Boston. This is Powell's last contribution to geology. He describes the theory of isostasy, which proposes that vertical movements in the crust of the earth are due to a process of compression and gravitational stress on flowing rock material that lies beneath the surface.

Warman separate not seen.

517."Intellections: A Psychological Study." *Open Court* 12 (1898): 641–652.

This essay begins with a lyrical narrative that traces Powell's thoughts as meander in stream-of-consciousness manner among observations, memories, and sensations. His purpose is to illustrate how association works and how concepts and symbols are formed. Sensations, reflections, perceptions, apprehensions, and ideations (thought) are all elements of judgment, which itself is one of five essential properties of matter.

Excerpted from *Truth and Error* (**#510**).

518."Interpretation of Folk-Lore." *Journal of American Folk-Lore* 8 (1898): 97–105.

This address was delivered in 1894 at a meeting of the American Folk-Lore Society. Powell revisits a number of his favorite themes, such as the five "Books of Humanities," the unity of the human race, and the evolution of the arts and industries. There are four different methods for the interpretation of facts found along the evolutionary path of folklore: imputation, personification, reification, and scientific.

519.["Remarks on Behalf of the Journal *Science*"]. *National Geographic Magazine* 9 (1898): 59–63.

Address at the Gardiner Greene Hubbard Memorial Meeting on 21 Jan. 1898. Hubbard and Alexander Graham Bell organized *Science* in 1883. By providing a central forum for news and discussion of scientific subjects, the journal significantly contributed to the organization of scientific research and cooperation among scientists.

520."Report of the Director of the Bureau of American Ethnology for the Year Ending June 30, 1896." *Annual Report of the Board of Regents of the Smithsonian Institution ... 1896*. 1898. Appendix 2, 32–45.

Largely the same as the Director's Report in the BAE's annual report for the year (**#522**).

521."Report of the Director of the Bureau of American Ethnology for the Year Ending June 30, 1897." *Annual Report of the Board of Regents of the Smithsonian Institution ... 1897*. Washington: GPO, 1898. Appendix 2, 33–45.

Largely the same as the Director's Report in the BAE's annual report for the year (**#527**).

522."Seventeenth Annual Report [of the Director] of the Bureau of American Ethnology." *Seventeenth Annual Report of the Bureau of American Ethnology to* *the Secretary of the Smithsonian Institution 1895–'96*. Pt. I. Washington: GPO, 1898. xxvii–lxxiii.

Powell once again addresses the five categories of human activities. These activities develop along an evolutionary path, leading ultimately to the integration of all humanity.

Warman separate not seen.

523."Whence Came the American Indians?" *Forum* 24 (1898): 676–688.

Reviewing the development of human activities among the American Indians according to his classification (art, industry, language, government, and philosophy), Powell concludes that mankind dispersed throughout the world before these activities developed. There is no evidence that the tribes "commingled" with Asians. Therefore, American Indians are indigenous. See also **#797**.

1899

Monographs

524.*Esthetology, or The Science of Activities Designed to Give Pleasure*. New York: Putnam's Sons, 1899.

This essay, along with "Philology" (**#535**), "Sociology" (**#525**), "Sophiology" (**#543**), and "Technology" (**#526**) from *American Anthropologist* constitute a body of essays intended to be published as the second of Powell's projected philosophical works, "Good and Evil." All bodies of the universe are judged to be good or evil based on the quality of their relations to humans (i.e., pleasure or pain). The resulting judgments give rise to human activities known as the arts, whose purpose is to secure pleasure. The science of the aesthetic arts classifies such activities into five types of pleasure: ambrosial, decorative, athletic, games, and fine arts. From the *American Anthropologist* (see **#528**). Also published in the 19th annual report of the BAE (**#537**).

525.*Sociology, or The Science of Institutions*. New York: Putnam's Sons, 1899.

Much of Powell's earlier writings on the evolution of institutions is incorporated in this essay. An institution is defined as a rule, or law, of conduct devised and agreed upon by men. Sociology is the science of the control of human activities by institutional devices. Its specialized fields include statistics, economics, civics, "historics" (history and biography) and ethics. Ethics is defined as "the science of conduct controlled by conscience" and incorporates religion. In the scientific stage of culture, the knowledge of good and evil depends upon the knowledge of truth and error. From the *American Anthropologist* (see **#531**). Also published in the 20th annual report of the BAE (**#558**).

526. *Technology, or The Science of Industries.* New York: Putnam's Sons, 1899.

Industry is an activity that is undertaken to promote the welfare of life. The science of industries is classified as substantiation (production of substances essential to life, such as agriculture), construction, mechanics, commerce, and medicine. Each class is likewise subdivided into five groups. A technologist is any man skilled in any one or a combination of these activities. From the *American Anthropologist* (see **#532**). Published in the 20th annual report of the BAE (**#560**).

Articles

527. "Eighteenth Annual Report [of the Director] of the Bureau of American Ethnology." *Eighteenth Annual Report of the Bureau of American Ethnology to the Secretary of the Smithsonian Institution, 1896–'97.* Pt. I. Washington: GPO, 1899. xxv–lvii.

Powell's report of activities is organized according to his five categories of "ethnic science" (esthetology, technology, sociology, philology, and sophiology). Also includes "List of Publications of the Bureau of American Ethnology" (lix–lxix).

Warman separate not seen.

528. "Esthetology, or The Science of Activities Designed to Give Pleasure." *American Anthropologist* ns 1 (1899): 1–40.

For annotation, see **#524**. See also **#534** and **#537**.

529. "Reply to Critics." *Science* ns 9 (1899): 259–263.

In response to critical reviews of *Truth and Error*, particularly by Lester Ward (**#879**), Powell defends his theory that there is in nature a self-organizing property, "affinity." Whether by consciousness or affinity, bodies (particles) are subject to this fifth essential constant of the universe.

530."Report of the Director of the Bureau of American Ethnology for the Year Ending June 30, 1898." *Annual Report of the Board of Regents of the Smithsonian Institution ... 1898*. Washington: GPO, 1899. Appendix 2, 36–48.

> Largely the same as the Director's Report in the BAE's annual report for the year (**#540**).

531."Sociology, or The Science of Institutions." *American Anthropologist* ns 1 (1899): 475–509, 695–745.

> For annotation, see **#525**. See also **#554** and **#558**.

532."Technology, or The Science of Industries." *American Anthropologist* ns 1 (1899): 319–349.

> For annotation, see **#526**. See also **#556** and **#560**.

1900

Monographs

533.*Archaeology; Commencement Address Delivered by Major J. W. Powell ... June 6, 1900*. Ed. Pauline M. Pitcher. Limestone College Studies, 1.6. Gaffney, SC: Limestone College Pr, 1900.

> This peculiar address is a combination of narrative, verse, and song. The narrative text is extracted from "The Five Books" (**#475**). Powell states that history formerly was written as it had been recorded by persons who witnessed an event and who therefore have a vested interest in the interpretation of the activities recorded. Historians now use their critical faculties and ancient history is thus informed by the science of archeology. The "books" are: Stone Book (geology), Ruin Book (ruins of habitation), and Tomb Book (burial structures and monuments).

534.*Esthetology, or The Science of Activities Designed to Give Pleasure*. Washington: GPO, 1900.

> For annotation, see **#524**. From the 19th BAE annual report (**#537**).

535. *Philology, or The Science of Activities Designed for Expression*. New York: Putnam's Sons, 1900.

> The fourth group of arts consists of languages, which are the artificial expression of concepts and designed to convey thoughts to others. The activities of philology are classified as emotional, oral, gestural,

written, and logistic languages. Language does not multiply by evolution but by integration and differentiation. From the *American Anthropologist* (**#541**). Also Published in 20th annual report of the BAE (**#557**).

Articles

536."Dualism Modernised." *Monist* 10 (1900): 383–396.

According to Powell, dualism is a universal belief of tribal society and most of modern mankind. Dualism holds that the world is composed of two types of bodies, material and ghostly. Dualism belongs to metaphysical philosophy. Modern attempts to develop a theory unifying the material and the spectral have failed because the mistake is made "of identifying force with will and making it the substrate of the world."

537."Esthetology, or The Science of Activities Designed to Give Pleasure." *Nineteenth Annual Report of the Bureau of American Ethnology to the Secretary of the Smithsonian Institution 1897–'98*. Pt. I. Washington: GPO, 1900. ix–xcii.

Contained in the "Director's Report" (**#540**). See **#524** for annotation. Also see **#528** and **#534**.

538."In Memoriam: Frank Hamilton Cushing; Remarks by J. W. Powell." *American Anthropologist* ns 2 (1900): 360–367.

Powell praises Cushing's skills in handicraft and his leadership as an ethnologist. "From the time we first went together to Zuni until the day of his death he was my companion and friend, and I loved him as a father loves his son."

539."The Lessons of Folklore." *American Anthropologist* ns 2 (1900): 1–36.

Using examples from primitive thought about causation and the organization of the natural world, such as ghosts, shamans, and dreams, Powell points out that similar magical thinking exists in modern systems of philosophy. Specifically, he continues his attack on metaphysics by criticizing James Ward's book on idealism, *Naturalism and Agnosticism*.

Warman separate not seen.

540."Nineteenth Annual Report [of the Director] of the Bureau of American Ethnology." *Nineteenth Annual Report of the Bureau of American Ethnology to* *the Secretary of the Smithsonian Institution 1897–'98*. Pt. I. Washington: GPO, 1900. xi–xcii.

Powell again reclassifies the entire work of the BAE according to his categories of human activity. He mentions his investigation of shell mounds on the Maine coast. In his comments on an accompanying paper by Jesse Fewkes, Powell recalls his travels in 1870 through the "Province of Tusayan" and remarks on the "fraternities" and social organization of the Hopi people (pp. xlvi–l). Contains "Esthetology" (**#534**, **#537**).

Warman separate not seen.

541."Philology, or The Science of Activities Designed for Expression." *American Anthropologist* ns 2 (1900): 603–637.

See **#535** for annotation. Also published in 20th annual report of the BAE (**#557**).

1901

Monographs

542.*The Categories*. New York: Putnam's Sons, 1901.

One of Powell's last and most difficult essays. He defines 25 categories of attributes that can be derived one from the other in a progressive pattern. There are five laws that accelerate evolution, two of which Powell claims he has previously demonstrated: the law of culture and "the instability of the heterogeneous" (in *Report on the Geology of the Eastern Portion of the Uintah Mountains,* **#249**). From *American Anthropologist* (**#545**).

543.*Sophiology, or The Science of Activities Designed to Give Instruction.* New York: Putnam's Sons, 1901.

The science of instruction is the fifth and last system of arts in Powell's pentalogical scheme. All but the last four pages are devoted to opinions, or philosophy. Philosophy has evolved from mythology to metaphysics to science. Science supplants superstition through verification. The art of sophiology classified instruction into five classes: nurture, oratory, education, publication, and research. From *American Anthropologist* (**#550**). Also published in the 20th annual report of the BAE (**#559**).

Articles

544."An American View of Totemism." *Man* [London] *2* (1902): 101–106.

A discussion of the distinctions between totemism and shamanism, as used in describing American and Australian tribal societies. See **#793** for responses.

545."The Categories." *American Anthropologist* ns 3 (1901): 404–430.

For annotation, see **#542**.

546."Classification of the Sciences." *American Anthropologist* ns 3 (1901): 601–605.

In 1883, Powell proposed to divide all science (knowledge) into three classes. He now proposes that five groups: nephelonomy (science of molecules), astronomy, geonomy (science of rocks), phytonomy (science of plants), and "zoönomy" (the science of animals). Each is subdivided into five yet-unnamed degrees of abstraction.

547. Introduction. *Zuñi Folk Tales.* By Frank Hamilton Cushing. New York, London: Putnam's Sons, 1901. vii–xvii.

Powell defines folklore as "the superstitions of the ignorant of today." He then explains the primitive thinking behind some of the myths and beliefs of "Amerinds" (a term newly adopted by Powell).

548."Report of the Director of the Bureau of American Ethnology for the Year Ending June 30, 1899." *Annual Report of the Board of Regents of the Smithsonian Institution ... 1899.* Washington: GPO, 1901. Appendix 2, 34–42.

Largely the same as the Director's Report in the BAE's annual report for the year (**#561**).

549."Report of the Director of the Bureau of American Ethnology for the Year Ending June 30, 1900." *Annual Report of the Board of Regents of the Smithsonian Institution ... 1900.* Washington: GPO, 1901. Appendix 2, 59–72.

Largely the same as the Director's Report in the BAE's annual report for the year (**#562**).

550."Sophiology, or The Science of Activities Designed to Give Instruction." *American Anthropologist* ns 3 (1901): 51–79.

For annotation, see **#543**. Also published in 20th annual report of the BAE (**#559**).

1902

Articles

551."The Scientific Explorer." *The Grand Canyon of Arizona; Being a Book of Words From Many Pens, About the Grand Canyon of the Colorado River in Arizona.* Chicago: Santa Fe Railway System, Passenger Dept., 1902. 18–32.

> Powell's last published essay was written for tourists. He reviews the history of Grand Canyon exploration and describes its physical features. There is an account of his meeting John D. Lee and Lee's wife at the mouth of the Paria River and an excerpt from *The Canyons of the Colorado*. This book by the Santa Fe is a larger edition of the pamphlet *Titan of Chasms* (**#552**) and contains additional articles by Hamlin Garland, Thomas Moran, and others. Later editions and reprints vary slightly.

552."The Scientific Explorer." *Titan of Chasms: The Grand Canyon of the Colorado*. Chicago: Santa Fe Railroad, 1902. 11–21.

> Same as **#551**. Powell's essay is one of three, with "Titan of Chasms" by Charles Higgins and "Greatest Thing in the World" by Charles Lummis. One of many editions and reprints published through 1918.

1903

Monographs

553.*Philology, or The Science of Activities Designed for Expression.* Washington: GPO, 1903.

> A pamphlet edition of the essay in the 20th annual report of the BAE (**#557**). For annotation, see **#535**.

554.*Sociology, or The Science of Institutions.* Washington: GPO, 1903.

> A pamphlet edition of the essay in the 20th annual report of the BAE (**#558**). For annotation, see **#525**.

555.*Sophiology, or The Science of Activities Designed to Give Instruction.* Washington: GPO, 1903.

> A pamphlet edition of the essay in the 20th annual report of the BAE (**#559**). For annotation, see **#543**.

556. *Technology, or The Science of Industries*. Washington: GPO, 1903.

A pamphlet edition of the essay in the 20th annual report of the BAE (**#560**). For annotation, see **#526**.

Articles

557. "Philology, or The Science of Activities Designed for Expression." *Twentieth Annual Report of the Bureau of American Ethnology to the Secretary of the Smithsonian Institution 1898–'99*. Washington: GPO, 1903. cxxxix–clxx.

For annotation, see **#535**. See also **#541** and **#553**.

558. "Sociology, or The Science of Institutions." *Twentieth Annual Report of the Bureau of American Ethnology to the Secretary of the Smithsonian Institution 1898–'99*. 1903. lix–cxxxviii.

For annotation, see **#525**. See also **#531** and **#554**.

559. "Sophiology, or The Science of Activities Designed to Give Instruction." *Twentieth Annual Report of the Bureau of American Ethnology to the Secretary of the Smithsonian Institution 1898–'99*. 1903. clxxi–cxcvii.

For annotation, see **#543**. See also **#550** and **#555**.

560. "Technology, or The Science of Industries." *Twentieth Annual Report of the Bureau of American Ethnology*. 1903. xxix–lvii.

For annotation, see **#526**. See also **#532** and **#556**.

561. "Twentieth Annual Report [of the Director] of the Bureau of American Ethnology." *Twentieth Annual Report of the Bureau of American Ethnology to the Secretary of the Smithsonian Institution 1898–'99*. Washington: GPO, 1903. vii–ccxxiv.

Contains "Technology" (**#560**), "Sociology" (**#558**), "Philology" (**#557**), and "Sophiology" (**#559**).

562. "Twenty-First Annual Report [of the Director] of the Bureau of American Ethnology." *Twenty-First Annual Report of the Bureau of American Ethnology to the Secretary of the Smithsonian Institution 1899–1900*. Washington: GPO, 1903. vii–xl.

Powell reports on investigations with Frank Cushing in Maine and a "reconnaissance" to Cuba and Jamaica with W. H. Holmes to trace cultural migration in the Western Hemisphere. He is resuming his "synthesis of the native American Languages," extending his classification to Mexico and Central America.

563."[Twenty-Second Annual Report] of the Director of the Bureau of American Ethnology." *Twenty-Second Annual Report of the Bureau of American Ethnology to the Secretary of the Smithsonian Institution 1900–1901*. Pt. 1. Washington: GPO, 1904. vii–xliv.

The work of WJ McGee occupies most of the director's report, in addition to the usual administrative reports. This and the following report (**#564**) were most likely written by McGee, who was acting director during Powell's illness.

564."Twenty-Third Annual Report [of the Director] of the Bureau of American Ethnology." *Twenty-Third Annual Report of the Bureau of American Ethnology to the Secretary of the Smithsonian Institution 1901–02*. Washington: GPO, 1904. vii–xlv.

The report refers to Powell's continued work on the classification of linguistic families of the Indians of Mexico and Central America, noting that its completion "was delayed by the illness of the Director during the later months of the year." This is the last annual report of the BAE under Powell's name.

1921

Articles

565. Address. *Proceedings and Debates of the Constitutional Convention; Held in the City of Helena, Montana, July 4th, 1889, August 12th, 1889*. Helena, MT: State Publishing Co, 1921. 820–823.

This speech was delivered to the Montana Constitutional Convention on 9 Aug. 1889. Powell, who was touring the West as a member of the Senate Special Committee on Irrigation, argues that Montana should adopt the drainage basin as the natural primary unit of government in the state. The farmer should own and control not only his own farmlands, "but also the lands where the timbers grow, and also the lands where the waters fall which make their lands valuable." In *Seeing Things Whole* (**#213**).

C. LATER REPRINTS OF WORKS BY POWELL

The following entries, arranged chronologically, comprise selected reprints of whole or large parts of whole works by Powell. The many 20th-century and early

21st-century works that contain short excerpts from *Exploration* (**#242**) and *Canyons of the Colorado* (**#466**) fall outside the scope of this bibliography. For excerpts and additional republications of the cited works below, refer to Earle Spamer's *Bibliography of the Grand Canyon and Lower Colorado River* (**#977**).

566. *First Through the Grand Canyon.* Ed. Horace Kephart. Intro. Horace Kephart. Outing Adventure Library. New York: Nelson Doubleday, 1915.

> The narrative part of the 1875 report republished without abridgement. Numerous editions and reprints of this work were published through 1940.

567. "Importance of Water Conservation." *Reclamation Era* 26.9 (1936): 201–202.

> Reprint of "Major Powell's Address" to the North Dakota Constitutional Convention (**#391**).

568. "Letters of Major J. W. Powell to the *Chicago Tribune*." *UHQ* 15 (1947): 73–88.

> Letters dated 24 May, 2 June, 7 June, 18 and 20 June, 23 June, 29 June, and 20 and 29 July 1869. See **#144**, **#220**, **#221**, **#222**, **#224**, and **#746**

569. "Major J. W. Powell's Report on His Explorations of the Rio Colorado in 1869." *UHQ* 15 (1947): 21–27.

> Reprinted from *New Tracks in North America* (London, 1870, 2nd ed. [559–564]). See **#227** and **#576**.

570. *The Exploration of the Colorado River.* Intro. Wallace Stegner. Chicago: U of Chicago Pr, 1957.

> Abridged from the 1875 publication. Stegner discusses the significance of Powell's exploration of the Colorado River and southern Utah, noting that he discovered the last unknown river (Escalante) and mountain range (Henry) in the continental United States. Illustrated with the original woodcuts.

571. *The Exploration of the Colorado River.* Natural History Library 11. Garden City, NY: Anchor Bks, 1961.

> Reprint of the narrative portion of the 1875 report. Published in cooperation with the American Museum of Natural History.

572.*The Exploration of the Colorado River and Its Canyons*. New York: Dover Publications, 1961.

This is an unabridged reprint of the 1895 title, *Canyons of the Colorado*.

573.*The Exploration of the Colorado River and Its Canyons*. New York: Doubleday, 1961.

Reprint of the narrative portion of the 1875 report.

574.*Report on the Lands of the Arid Region of the United States, With a More Detailed Account of the Lands of Utah*. Ed. Wallace Stegner. Cambridge, MA: Belknap Pr, 1962.

Reprint of the complete text of the second edition (1879), including maps. With introduction and notes by Stegner and a chronology of Powell's life.

575.*Canyons of the Colorado*. New York: Argosy-Antiquarian Ltd., 1964.

Facsimile of the 1895 publication.

576."Major J. W. Powell's Report on His Explorations of the Rio Colorado in 1869." *New Tracks in North America*. By William A. Bell. Albuquerque, NM: Horn and Wallace, 1965. Appendix D, 559–564.

Facsimile of the original 1870 publication (**#227**, **#569**).

577.*Indian Linguistic Families of America North of Mexico*. Ed. Preston Holder. Bison Book BB301. Lincoln: U of Nebraska Pr, 1966. 83–221.

Originally published in the BAE 7th annual report (**#422**). This book also contains a reprint of Franz Boas's introduction to *Handbook of American Indian Languages* (**#798**).

578.*Down the Colorado: Diary of the First Trip Through the Grand Canyon, 1869*. Intro. and foreward Don D. Fowler. Illus. Eliot Porter. New York: E. P. Dutton, 1969.

An abridgement of Part I of *Exploration*. This oversize book is richly illustrated with photographs by Eliot Porter. The introduction by Fowler provides historical background for *Exploration*. An epilogue by photographer Porter reflects on what was lost with the erection of the Glen Canyon dam. Porter's photographs are stunning. Republished several times (Promontory Pr, Arrowood Pr, and Stanford U) and an excerpt was published under same title in *American Heritage*, v. 20, June 1969 (pp. 52–61, 83).

579. Foreword. *Picture-Writing of the American Indians* Vol. 1. Garrick Mallery. New York: Dover, 1972. v–x.

This book is an unabridged republication of a paper from the BAE 10th annual report. The foreword by Powell is a reprint from the Director's Report (**#450**). Powell comments on the illustrations and the organization of the report.

580.*The Hopi Villages: The Ancient Province of Tusayan*. 1875. Intro. Lollie W. Campbell. Wild and Woolly West Books 21. Palmer Lake, CO: Filter Pr, 1972.

Reprint of "The Ancient Province of Tusayan" (**#243**) with additional photographs.

581.*An Overland Trip to the Grand Canyon*. Wild and Woolly West Books 28. Palmer Lake, CO: Filter Pr, 1974.

Reprint of the 1875 article in *Scribner's* (**#246**). Text and woodcuts from Powell's plus drawings from other sources.

582. Letter. *The National Elgin Watch Company Illustrated Almanac 1871*. Northbrook, IL: Bernard J. Edwards, 1977. 19.

Facsimile of the original (**#229**) with an additional cover.

583."Statement of the Director of the Geological Survey [on the Organization of Scientific Work in the General Government]." *Testimony Before the Joint Commission to Consider the Present Organization of the Signal Service, Geological Survey, Coast and Geodetic Survey, and the Hydrographic Office of the Navy Department, With a View to Secure Greater Efficiency and Economy of Operation*. New York: Arno, 1980.

Reprint of original (**#843**) in two volumes.

584.*The Cañons of the Colorado: The 1869 Discovery Voyage Down the Colorado River From Wyoming Into Colorado, Through Utah, to Arizona and Nevada*. Golden, CO: Outbooks, 1981.

Facsimile of the 1875 Scribners articles (**#243**, **#244**, **#246**) with editor's preface and maps.

585.*La Exploracion del Gran Cañon Del Colorado*. Trans. Marta Perez. Nan Shan 8. Barcelona: Laertes S. A. de Ediciones, 1983.

Spanish translation of the 1895 *Canyons of the Colorado*.

586.*Report on the Lands of the Arid Region of the United States, With a More Detailed Account of the Lands of Utah.* Intro. T. H. Watkins. Boston, MA: Harvard Common Pr, 1983.

Facsimile of the 1879 edition.

587.*The Exploration of the Colorado River and Its Canyons.* Intro. Wallace Stegner. Penguin Nature Library. New York: Penguin Books, 1987.

Reprint of the 1875 report.

588.*Les Canyons du Colorado.* Trans. Gail Wagman and Marc de Gouvenain. Terres D'Aventure. Arles, France: Actes Sud, 1995.

French adaptation of the 1895 *Canyons of the Colorado.*

589.*Canyons of the Colorado.* Ed. and illus. Joseph Holmes. Foreword David R. Brower. San Francisco: Chronicle Bks, 1996.

Oversize picture book. Photographs and essays by Joseph Holmes interspersed with text excerpted from the 1875 *Exploration.*

590.*Introduction to the Study of Indian Languages; Indian Linguistic Families of America.* Origins of American Linguistics, 1643–1914, 8. London: Routledge/Thoemmes, 1997.

Reproduction of both titles (see **#276**, **#422**), published in a single volume without the plates or maps.

591.*Höllenfahrt durch den Grand Canyon; Expeditionsbericht über die erste Befahrung und Erforschung des Green River und des Colorado River in 1869.* Oberschleissheim: Pollner Verlag, 2001.

Roughly translated: "Hell-trip through the Grand Canyon." German adaptation of *Exploration.*

592.*Exploration of the Colorado River and Its Canyons.* Intro. Anthony Brandt. Washington: Adventure Classics/National Geographic, 2002.

Reprint of the 1895 *Canyons of the Colorado.*

Chapter 3

Works About John Wesley Powell

A. BIOGRAPHICAL PUBLICATIONS, GENERAL AND EARLY YEARS

1. Books and Articles

593. Andrew, Bunyan, and Richard E. Rann. "John Wesley Powell: Wesleyan's First Scientific Impulse." *Illinois Wesleyan University Bulletin* 67.5 (1969): 4–5, 15.

> Brief synopsis of Powell's tenure at IWU and a history and description of the Powell Collection of Indian pottery.

594. Aton, James M. "Powell, John Wesley." *Nineteenth-Century American Western Writers*. Ed. Robert L. Gale. Dictionary of Literary Biography 186. Detroit: Gale Research, 1997. 296–305.

> Biography, overview of Powell's published works, and a selected bibliography.

595. Baars, D. L. "Major John Wesley Powell, Colorado River Pioneer." *Geology and Natural History of the Fifth Field Conference, Powell Centennial River Expedition 1969*. Durango, CO: Four Corners Geological Society, 1969. 10–18.

> Baars relies heavily on Darrah for this biography. Includes quotations from *Exploration.* See **#924**.

596. Baker, Marcus. "Major J. W. Powell; Personal Reminiscences of One of His Staff." *Open Court* 17 (1903): 348–351.

> Powell is compared to Lincoln for his open-door policy with his Geological Survey associates and employees. Baker recalls that he

would read poetry and Ruskin's essays to Powell during sittings for a bust. See also **#602**, **#621**, **#622**, and **#645–#648**.

597. Bear, Elwood. "John Wesley Powell—He Tamed the Colorado." *Civil Service Journal* 9.3 (1969): 2–7.

Uncritical overview of Powell's first expedition and subsequent government career.

598. Brewer, William H. "John Wesley Powell." *American Journal of Science* 4th ser. 14 (1902): 377–382.

Brewer recalls an after-dinner conversation with Powell about why Powell wasn't afraid of encountering a large waterfall in the Grand Canyon.

599. Bryant, H. C. "John Wesley Powell." *Cosmos Club Bulletin* 1.8 (1948): 2–5.

A report on the dedication of a memorial monument to Powell at the Grand Canyon in 1918.

600. Cantelon, Philip L. *John Wesley Powell Didn't Throw in the Towel.* Philosophical Society of Washington. 28 January 2002. <www.philsoc.org/2000Fall>.

Web publication of an address to the PSW at the Cosmos Club on 5 Jan. 2001. Cantelon reviews the history of the Cosmos Club and the PSW and Powell's role in their establishment.

601. Carlson, Curt. "Scholar From the Past: John Wesley Powell." *Illinois State Scholar* 5 (Spring–Summer 1992): 24–25.

Summary of Powell's accomplishments and his career at ISU.

602. [Carus, Paul]. "Major Powell, the Chief." *Open Court* 16 (1902): 716.

Open Court's editor praises Powell as a born leader. Carus quotes from the text of the diploma that Powell received from the University of Heidelberg, awarded for his work as chief of the USGS. See also **#596**, **#621**, **#622**, and **#645–#648**.

603. Chamberlain, Alexander F. "In Memoriam: John Wesley Powell, 1834–1902." *Journal of American Folklore* 15 (1902): 199–204.

A memorial tribute, including bibliography. There are factual errors.

604. Cole, Charles E. "John Wesley Powell (1834–1902)." *Something More Than Human; Biographies of Leaders in American Methodist Higher Education.* Nashville, TN: United Methodist Board of Higher Education and Ministry, 1986. 185–202.

The emphasis in this biography is on Powell's religious formation and his contribution to higher education through his fieldwork and scientific organizations.

605. Crossette, George. *Founders of the Cosmos Club of Washington, 1878: A Collection of Biographical Sketches and Likenesses of the Sixty Founders.* Washington: Cosmos Club, 1966. 130–132.

Biographical sketch.

606. Dall, William Healey. "John Wesley Powell, 1834–1902." *Bulletin of the Philosophical Society of Washington, 1900–1904* 14 (1904): 300–308.

Eulogy read before the Society on 21 May 1904.

607. Darrah, William Culp. "John Wesley Powell and an Understanding of the West." *UHQ* 37 (1969): 146–151.

Darrah praises Powell for his role in environmental conservation.

608.——. "Major John Wesley Powell, 1834–1902." *UHQ* 15 (1947): 19–20.

Brief biography.

609.——. *Powell of the Colorado.* Princeton, NJ: Princeton U Pr, 1951.

Prior to *A River Running West,* Donald Worster's biography of Powell (**#689**), this biography was the best book-length biography of the *whole* of Powell's life. Stegner's study (**#671**) is primarily concerned with Powell's work in the West, and Terrell's book (**#676**) lacks substance. Darrah was also the first biographer to document his sources for Powell's early years through the Civil War. He writes uncritically of Powell as a heroic explorer and proponent of reclamation projects. Republished by Princeton in paperback (1969).

610.——. "Powell of the Colorado." *UHQ* 28 (1960): 223–231.

Darrah states that Powell, more than any other person, was responsible for bringing about reforms in federal land policies. He chose the federal government as his place to work and the public domain as the object of his study and stewardship.

611. Davis, William Morris. "John Wesley Powell." *National Academy of Sciences Biographical Memoirs*. Vol. 8. Washington: Judd & Detweiler, 1915. 11–83.

Presented to the Academy in 1913. Davis focuses on Powell's publications to review Powell's work and activities. He critiques Powell's writing style and concentrates on his accomplishments in the larger scientific community. This biography is a standard source for subsequent biographers.

612. Dellenbaugh, Frederick Samuel. *John Wesley Powell, a Brief Review of His Career*. New York: G. P. Putnam's Sons, 1903.

This is an offprint of the epilogue in *The Romance of the Colorado River* (pp. 371–386; **#109**).

613.——. "Memorial to John Wesley Powell." *American Anthropologist* ns 20 (1918): 432–436.

Dellenbaugh describes the planning, design, and dedication of the Powell monument in the Grand Canyon that commemorates the 1869 voyage. Dellenbaugh played a major role in the controversial omission of the Howland brothers and Dunn from the monument's plaque.

614. Dorman, Robert L. *A Word for Nature: Four Pioneering Environmental Advocates, 1845–1913*. Chapel Hill: U of North Carolina Pr, 1998.

Powell keeps company in this volume with George Perkins Marsh, Henry David Thoreau, and John Muir. In Powell's biography (pp. 173–218), Dorman emphasizes Powell's geological fieldwork, the intellectual foundations of his views on the development of the West, and his political struggles in the late 1880s.

615. Elman, Robert. "John Wesley Powell." *First in the Field: America's Pioneering Naturalists*. New York: Mason/Charter, 1977. 157–199.

General biography.

616. Fenton, Carroll Lane, and Mildred Adams Fenton. "Canyon's Conqueror." *The Story of the Great Geologists*. Garden City, NY: Doubleday, Doran and Co, 1945. 232–250

The chapter on Powell emphasizes his work in geology.

617. Flack, J. Kirkpatrick. *Desideratum in Washington; The Intellectual Community in the Capital City, 1870–1900*. Cambridge, MA: Schenkman Pub Co, 1975.

An overview of the personalities, associations, and societies of late 19th-century Washington. Powell was a member of many of the newly formed organizations, composed of individuals involved in the city's public and intellectual life. Flack looks at the relationship of the Cosmos Club to the consolidation of the Western surveys. Contains a useful "Essay on Sources" (pp. 171–184).

618. Gilbert, Grove Karl. "John Wesley Powell." *Science* 16 (1902): 561–567.

An obituary notice by one of Powell's closest friends. Gilbert notes that Powell's personal work is not easily separated from the work accomplished through the organization of others. See also **#620**.

619.——. "John Wesley Powell." *Geographical Journal* 20 (1902): 663–664.

Extracted from obituary notice in *Science* (**#618**).

620.——. "John Wesley Powell." *Annual Report of the Board of Regents of the Smithsonian Institution ... 1902*. (1903): 663–640.

Revised by the author from an article published in *Science* (**#618**). Offprint published with separate cover (1903).

621.——. "John Wesley Powell, Pt. V: The Investigator." *Open Court* 17 (1903): 228–239, 281–290.

Gilbert reviews Powell's published contributions to geology, anthropology, and the philosophy of science. Like other "great investigators," Powell began with observation and collection of facts, then discussed his own observations (hypotheses), and finally created theories based on his observations. One of a six-part series (see **#602, #622, #645–#648**).

622.——. "John Wesley Powell, Pt. VI: The Promoter of Research." *Open Court* 17 (1903): 342–347.

As "a man of affairs," Powell ably organized and strengthened the BAE and the USGS. He also was key to establishing Washington as a scientific center of the United States. One of a six-part series (see **#602, #621, #645–#648**).

623.——. "Powell as a Geologist." *Proceedings of the Washington Academy of Sciences* 5 (1903): 113–117.

Gilbert reviews Powell's contribution to applied science and praises him as the director of the USGS. In "John Wesley Powell: Proceedings of a Meeting...." See **#640**.

624. Gilbert, Grove Karl, ed. *John Wesley Powell: A Memorial to an American Explorer and Scholar; Comprising Articles by Mrs. M. D. Lincoln (Bessie Beach), Grove Karl Gilbert, Marcus Baker, and Paul Carus*. Chicago: Open Court, 1903.

Articles reprinted from *Open Court* with slight revision. See **#602**, **#621**, **#622**, and **#645–#648**.

625.[Grosvenor, Gilbert H.]. "John Wesley Powell." *National Geographic Magazine* 13 (1902): 392–394.

Brief obituary with a full-page portrait, signed G. H. G.

626. Guziak, Margaret Melloy. "The Enduring Saga of John Wesley Powell." *Indian Trader* 28.8 (August 1997): 24–26.

Sketch of Powell and the Colorado River expeditions.

627. Henderson, David B. "Powell as a Soldier." *Proceedings of the Washington Academy of Sciences* 5 (1903): 100–105.

In this review of Powell's military career during the Civil War, Henderson describes Powell as a "tender, sympathetic man." In "John Wesley Powell: Proceedings of a Meeting ..." (**#640**).

628. Hendrickson, Walter B. "The Forerunners of the Illinois State Academy of Science." *Transactions of the Illinois State Academy of Science* 56 (1963): 105–128.

The Illinois Natural History Society is one of the forerunners discussed in this article. Powell organized a local chapter of the Society in Bloomington and was successful in securing state funding to create and curate a state museum in Normal. However, his self-interest contributed to the demise of the Society in 1871.

629.——. "Nineteenth-Century Natural History Organizations in Illinois." *Journal of the Illinois State Historical Society* 54 (1961): 246–267.

A overview of the history of the Illinois State Academy of Science, including Powell's role as organizer of the Bloomington chapter of the

Illinois Natural History Society and curator of its museum in Normal (1859–1871). As curator, Powell neglected science in Illinois.

630. Hobbs, William Herbert. "John Wesley Powell, 1834–1902." *Scientific Monthly* 39 (1934): 519–529.

This biography is illustrated with photographs of various Powell memorial monuments.

631. Holmes, W. H. "John Wesley Powell." *Twenty-Fourth Annual Report of the Bureau of American Ethnology, 1902–03*. Washington: GPO, 1907. xxxv–xxxvii.

A "necrology." Holmes observes that the publications of the BAE are a "splendid monument to his memory."

632. Horton, Peggy S. "The Ohio Years of Major John Wesley Powell." *Ohioana of Ohio and Ohioans* (Spring 1964): 18–20.

The author relies on Darrah for an account of the Powell family in Jackson, Ohio, and young John's relationship with George Crookham. Horton recalls the dedication of Jackson's memorial to Powell in 1934.

633. Howard, Robert West. "John Wesley Powell—Conqueror of the Colorado." *Conservationist* 32 (November–December 1978): 17–19, 48.

Brief biography.

634.——. "Powell of the Genessee." *Westerners, New York Posse Brand Book* 16.2 (1969): 25–29, 32.

The author suggests that the topography of upstate New York deeply affected Powell's career in natural science.

635. Hursch, Carolyn J. "Mapping the Colorado." *American History* 31.3 (1996): 34–40, 58–60.

Biography with emphasis on the 1869 expedition.

636.*Illinois at Vicksburg*. Chicago: Illinois-Vicksburg Military Park Commission, 1907.

Detailed information and photographs of the regimental monuments and markers and the Illinois state monument erected in the Vicksburg National Military Park. Includes an account of Powell and his unit, Battery F, Second Artillery, of the Illinois Volunteers.

637."In Memory of John Wesley Powell." *Science* ns 16 (1902): 782–790.

Samuel Langley transmits to *Science* "minutes of a meeting held at the USNM preceding the funeral of Major J. W. Powell, Director of the Bureau of American Ethnology, Sept. 26, 1902." The eulogies, introduced by Richard Rathbun, are by William Dall, Daniel Gilman, Charles Walcott, William Harris, Marcus Baker, and WJ McGee.

638. James, Preston E. "John Wesley Powell, 1834–1902." *Geographers: Bibliographical Studies* 3 (1979): 117–124.

An assessment of Powell's work in geography and ethnology, and the influence of his ideas through his work in government science. James gives Powell credit for the conservation movement in the United States. Includes bibliography and detailed chronology.

639."John Wesley Powell." *American Anthropologist* ns 4 (1902): 564–565.

Obituary notice. Reprinted in **#801**.

640."John Wesley Powell: Proceedings of a Meeting Commemorative of His Distinguished Services, Held in Columbian University Under the Auspices of the Washington Academy of Sciences, February 16, 1903." *Proceedings of the Washington Academy of Sciences* 5 (1903): 99–187.

Contents include eulogies by Charles D. Walcott (**#680**), David B. Henderson (**#627**), Charles R. Van Hise (**#677**), Grove K. Gilbert (**#623**), WJ McGee (**#653**) and Samuel P. Langley (**#643**); and "Catalogue of the Published Writings of John Wesley Powell" by P. C. Warman (**#979**). Published also as a separate pamphlet (1903).

641. Jorgenson, Pat. "True Grit Geologist: The Vision of John Wesley Powell." *People, Land and Water* (U.S. Dept. of the Interior) 6.2 (1999): 66–68.

Overview of Powell's life and work.

642. Kolb, Anne. "John Wesley Powell (1834–1902)." *Backpacker* 2.1 (1974): 63–[65], 90, 94, 97.

A laudatory biography of Powell, who was a victim of "backstabbing" by his congressional enemies.

643. Langley, Samuel P. "Powell as a Man." *Proceedings of the Washington Academy of Sciences* 5 (1903): 127–130.

Powell had "an imagination akin to that of the poet." As a friend, he was sincere, truthful, and steadfast; as a man he was brave, generous, and stoic. In "John Wesley Powell: Proceedings of a Meeting ..." (**#640**).

644.——. "Bureau of American Ethnology." *Annual Report of the Board of Regents of the Smithsonian Institution ... 1903*. Washington: GPO, 1904. 18–20.

Tribute to Powell in the "Report of the Secretary."

645. Lincoln, Martha D. "John Wesley Powell, Pt. I: Boyhood and Youth." *Open Court* 16 (1902): 705–715.

The first of four articles by Mrs. M. D. Lincoln (penname "Bessie Beech") in a six-part serialized biography and assessment of Powell's career. Based on interviews by Lincoln with Powell during 1885–1895, this article became the principal source of information concerning Powell's childhood and youth for future biographers. See also **#602**, **#621**, **#622**, and **#646–#648**.

646.——. "John Wesley Powell, Pt. II: The Soldier." *Open Court* 17 (1903): 14–25.

Powell's Civil War years, 1861–1865. Contains some factual errors. One in a six-part series (see **#602**, **#621**, **#622**, **#645**, **#647**, **#648**).

647.——. "John Wesley Powell, Pt. III: The Professor." *Open Court* 17 (1903): 86–94.

Powell's post–Civil War years (1865–1868) as a professor at IWU and ISU, including the field trips to the Colorado Rocky Mountains. One in a six-part series (see **#602**, **#621**, **#622**, **#645**, **#646**, **#648**).

648.——. "John Wesley Powell, Pt. IV: The Explorer." *Open Court* 17 (1903): 162–174.

Primarily passages excerpted from Powell's narrative in *Exploration*. One in a six-part series (see **#602**, **#621**, **#622**, **#645–#647**).

649."Major John W. Powell." *American Antiquarian and Oriental Journal* 14.3 (1892): 155–158.

"This sketch of Maj. Powell is one of a series ... [of] the leading writers in archaeology." The text of this anonymous article has numerous similarities to Lincoln's articles (**#645–#648**).

650."Major John Wesley Powell." *Illinois Wesleyan Magazine* 7.3 (1902): 127–129.

Obituary taken from the Bloomington (Illinois) *Daily Pantagraph* of 24 Sept. 1902. The text is similar to Wilder (**#688**).

651."Major John Wesley Powell Dead: A Famous Scientist and a Pioneer in the Irrigation Movement." *Forestry and Irrigation* 8 (1902): 398–400.

Obituary.

652. Marshall, Helen E. *Grandest of Enterprises: Illinois State Normal University 1857–1957.* Normal: Illinois State Normal U, 1956.

Examines Powell's roles at ISU as a professor, leader of scientific expeditions in 1867 and 1868, and curator of the Illinois State Natural History Museum (pp. 117–130).

653. McGee, WJ. "Powell as an Anthropologist." *Proceedings of the Washington Academy of Sciences* 5 (1903): 118–127.

Powell became his best as an anthropologist, according to one of his closest associates. He began his career as a Linnean, then became a Darwinian, and finally a Baconian. McGee presents Powell's plan for three works (*Truth and Error* was one; see **#510**), which would frame a "cosmic compendium." In "John Wesley Powell: Proceedings of a Meeting ..." (**#640**).

654. Meadows, Paul. *John Wesley Powell, Frontiersman of Science.* U of Nebraska Studies; New Ser. 10. Lincoln: U of Nebraska, 1952.

In 1952, Meadows could accurately refer to Powell "as an almost forgotten figure." Organized according to Powell's principal areas of work and interest. Meadows studies Powell's intellectual life, primarily through his writings and those of his contemporaries. There are extensive endnotes and a good bibliography of Powell's publications.

655."Memorial to John Wesley Powell." *Science* 43 (1916): 16.

About the monument built on the South Rim.

656. Merrill, George P. "John Wesley Powell." *American Geologist* 31 (1903): 327–333.

Review of Powell's work in geology. Argues that Powell's greater success was as an administrator rather than as a scientist. Also published as a pamphlet in 1903.

657. Miller, Peter, and Bruce Dale. "John Wesley Powell: Vision for the West." *National Geographic* 185.4 (April 1994): [86–115].

This biography is also a photo essay of places from Powell's life as they appear today. Powell laid the foundation for federal land policy, leaving the nation a rich legacy.

658."A Monument to J. W. Powell." *Science* ns 20 (1904): 854.

Report of an excursion by the 8th International Congress of Geologists to the Grand Canyon, where a memorial meeting was held and a committee established to request that a monument be erected.

659. Moring, John. "John Wesley Powell." *Men With Sand: Great Explorers of the North American West*. Helena, MT: TwoDot, 1998. 185–199.

Collection of biographies, including Powell's.

660. Morris, Lindsey G. "John Wesley Powell: Scientist and Educator." *Illinois State University Journal* 31.3 (1969): 3–46.

This biography is an abridgement of Morris's master's thesis (**#987**). There are errors of fact in the text and bibliography, but Morris was one of the first biographers to examine sources from Powell's years in Illinois.

661. Muench, Joyce Rockwood. "John Wesley Powell." *Arizona Highways* 41.4 (April 1965): 4–8.

The usual biography.

662. Myres, John L. "John Wesley Powell: Born March 24, 1834; Died September 23, 1902." *Man* 3.10 (1903): 23–25.

Obituary.

663. Page, Jake. "John Wesley Powell: Champion of the West." *Destination Discovery* 11.2 (1995): 32–37.

A highly laudatory biography ("He was a prophet to his people").

664. Rabbitt, Mary C. "John Wesley Powell, His Life and Times." *Geotimes* 14 (1969): 10–12.

Brief biography.

665.[Rabbitt, Mary C.]. *John Wesley Powell: Soldier, Explorer, Scientist.* Washington: U.S. Dept. of the Interior, Geological Survey, 1969.

Pamphlet biography of Powell with an emphasis on his work as a geologist and director of the USGS. Unsigned. In 1976 and 1980, this was slightly revised and reprinted with authorship credited to Rabbitt.

666.[Rabbitt, Mary C.]. "John Wesley Powell: Soldier, Explorer Scientist." *Greater Llano Estacado Southwest Heritage* 4.4 (1974–1975): 7–15.

Same text as pages 5–23 of **#665**. Unsigned.

667. Roscoe, Ernest J. "John Wesley Powell—Pioneer Illinois Malacologist." *Annual Report of the American Malacological Union* (1961): 2.

Abstract of an address to the Union in which Roscoe discusses Powell's involvement with the Illinois Natural History Society as Curator of Conchology.

668.——. "Major John Wesley Powell, Illinois Naturalist." *Bulletin of the Field Museum of Natural History* 40.5 (1969): 11–14.

Summarizes Darrah's and Stegner's accounts of Powell's youth as a naturalist, soldier, and teacher in Illinois.

669. Ross, Kip. "Fateful Meeting, Fateful Year 1878." *Cosmos Club Bulletin* 11.10 (1958): 3–12.

Ross examines key events in Washington in 1878, when the city was the nation's scientific center. Among important events are the founding of the Cosmos Club by Powell, the publication of *Report on the Arid Lands*, and the recommendation to Congress by the National Academy of Science to consolidate the four Western surveys.

670. Spitzka, Edward Anthony. "A Study of the Brain of the Late Major J. W. Powell." *American Anthropologist* ns 5 (1903): 585–643.

Among scientists and intellectuals of this era, it was not uncommon to leave their brains for study and measurement. A roster was kept, which ranked brains according to weight, based on the idea that brain weight correlates with intelligence. Powell and WJ McGee had laid a wager on whose brain was larger. Spitzka correlates Powell's mental characteristics with the anatomy of his brain. Of particular interest are the remembrances of many of Powell's friends that Spitzka collected for his study. Also published in pamphlet form.

671. Stegner, Wallace Earle. *Beyond the Hundredth Meridian: John Wesley Powell and the Second Opening of the West.* Boston: Houghton Mifflin, 1954. [Rpt. Lincoln: U of Nebraska Pr, 1982; New York: Penguin Bks, 1992.]

> Introduction by Bernard DeVoto. This book-length biography portrays Powell as a preservationist, environmental champion, and agent of social change. Stegner is primarily concerned with Powell's career in Washington as an incorruptible government scientist and a visionary regarding public land policy for the arid lands of the American West. Extensive notes. An essential work, along with Darrah (**#609**) and Worster (**#689**).

672.——. "A Dedication to the Memory of John Wesley Powell, 1834–1902." *Arizona and the West* 4 (1962): 1–4.

> In this review of Powell's career, Stegner states that Powell's greatest contribution was the formulation of institutions appropriate to Western conditions. Beyond his own efforts, Powell suggested work to others and supported them work through his bureaus. Includes a bibliography of principal publications by Powell.

673. Steinbacher-Kemp, William. "John Wesley Powell: How a One-Armed Naturalist From Bloomington Became the Nation's Greatest Explorer." *Illinois Heritage* 5 (Fall 2001): 7–12.

> The author argues that Powell's early years in the Midwest formed Powell's character and significantly influenced his attitude toward the land.

674. Stirling, Matthew W. "John Wesley Powell Re-Discovered." *Cosmos Club Bulletin* 12.1 (1959): 2–9.

> An address presented on the occasion of the 80th anniversary of the Cosmos Club.

675. Strong, Douglas Hillman. "John Wesley Powell." *Dreamers & Defenders: American Conservationists*. Lincoln, NE: U of Nebraska Pr, 1988. 39–59.

> This biographical essay focuses on Powell's geological surveys and his proposals for land reform in the arid West. *Dreamers and Defenders* is a revised edition of *The Conservationists* (1971).

676. Terrell, John Upton. *The Man Who Rediscovered America: A Biography of John Wesley Powell.* New York: Weybright and Talley, 1969.

> A pedestrian book-length biography that relies on secondary sources.

677. Van Hise, Charles Richard. "Powell As an Explorer." *Proceedings of the Washington Academy of Sciences* 5 (1903): 105–112.

> Powell is compared to Fritjof Nansen and other leaders of men who plan carefully, are concerned with the safety of their men, and explore for the purpose of advancing objective knowledge and the philosophy of science. In "John Wesley Powell: Proceedings of a Meeting …" (**#640**).

678. Vance, Joseph W., ed. *Report of the Adjutant General of the State of Illinois; Containing Reports for the Years 1861–'66.* Rev. ed. Springfield, IL: H. W. Rokker, 1886.

> V. 8 contains an organizational record of Powell's army unit, the Illinois Second Regiment, Battery F, and an account of Powell's wounding at the Battle of Shiloh in 1862.

679. Walcott, Charles D. "John Wesley Powell." *Twenty-Fourth Annual Report of the Geological Survey*. Washington: GPO, 1907. 271–287.

> In his eulogy, Walcott quotes heavily from other sources: Dellenbaugh's epilogue in *Romance of the Colorado River* (**#109**); Gilbert's tribute in *Science* (**#618**); and McGee's address at the memorial meeting at the Washington Academy of Science (**#653**). Also cited as "Necrology of John Wesley Powell."

680. ——. "Remarks by Charles D. Walcott." *Proceedings of the Washington Academy of Sciences* 5 (1903): 99–100.

> An assessment of Powell as an administrator. Powell possessed insight, scientific imagination, and initiative. In "John Wesley Powell: Proceedings of a Meeting …" (**#640**).

681. *War of the Rebellion: A Compilation of the Official Record of the Union and Confederate Armies*. Washington: GPO, 1880–1901.

> This 128-volume set is essential to the study of the Civil War. The work of Powell and his men is detailed in Series I, v. 24, pt. 2 (pp. 197–203). Other accounts of operations in which Powell took part are in Series I, v. 49, pt. 2 (pp. 774–780); Series 1, v. 31, pt. 3 (pp. 564–570); and Series 1, v. 32, pt. 1 (pp. 164–171).

682. Ward, Lester Frank. "Professor John W. Powell." *Popular Science Monthly* 20 (1882): 390–397.

> In this biography of his good friend, Ward describes Powell as fitting the "pattern of the American self-made man." Powell is a frank, genial

man who strongly favors persons "of independent thought." Ward discusses the consolidation of the four Western surveys and highlights Powell's philosophy of science. Reprinted in *Glimpses of the Cosmos*, v. 2 (**#683**).

683.——. "Sketch of Professor John W. Powell (With Portrait)." *Glimpses of the Cosmos*. Vol. 2. New York: Putnam's Sons, 1913. 426–438.

In preparation for the article in *Popular Science* (**#682**), Ward conducted a long interview with Powell in May 1881. The entire article is reprinted here.

684. Warren, George A. "John Wesley Powell; a Tribute to a Great Bureau Chief and Man of Science...." *Federal Employee* 3 (1918): 763–765.

A report on the erection and dedication of a monument to Powell in the Grand Canyon.

685. Washburn, Wilcomb E. *The Cosmos Club of Washington; a Centennial History 1878–1978*. Washington: Cosmos Club, 1978.

Credit is given Powell as the founder of the Cosmos Club, with accounts of his activities a member. Includes a historically inaccurate "vignette" by H. C. Bryant (pp. 286–288).

686. Wetmore, George Peabody. "Memorial to John Wesley Powell." 60th Cong., 1st Sess., S. Rept. 293 Serial 5218. Washington: GPO, 1908.

This report from the Senate Committee on the Library, accompanied a bill to erect a memorial to Powell on the brink of the Grand Canyon, contains a sketch of Powell's life and work.

687. Widdison, Jerrold G. "John Wesley Powell." *Arizona Highways* 45.3 (March 1969): 4–7.

The author identifies Powell's greatest accomplishment as "the way he combined learning with action, so that his learning mattered."

688. Wilder, William H. "John Wesley Powell, Professor, 1865–1868." *An Historical Sketch of the Illinois Wesleyan University, Together With a Record of the Alumni. 1857–1895*. Bloomington: Illinois Wesleyan U Pr, 1895. 86–88.

Biographical sketch.

689. Worster, Donald. *A River Running West: The Life of John Wesley Powell.* New York: Oxford U Pr, 2001.

This scholarly biography fully documents and interprets the whole of Powell's life and work. Worster recognizes Powell's complex and sometimes contradictory character. His life is detailed within the context of the politics, social milieu, intellectual life, and personalities of the last half of the 19th century. Though his admiration for Powell is clear, Worster recognizes his subject's shortcomings and failures. Beyond his achievements as an explorer, scientist, and bureaucrat, Powell's significance is seen as a witness to and participant in the great religious and social changes that occurred as the American nation matured. Chapter notes and accompanying bibliography provide excellent resources for students and researchers.

2. Encyclopedias

690. Axelrod, Alan. "Powell, John Wesley." *Encyclopedia of the American West.* Vol. 3. Ed. Alan Axelrod and Charles Phillips. New York: Macmillan, 1996. 1320–1321.

691. [Benjamin, Marcus]. "Powell, John Wesley." *Appleton's Cyclopaedia of American Biography.* Vol. 5. New York: D. Appleton, 1888. 94–95.

Unsigned. See **#14**.

692. Lee, Lawrence B. "Powell, John Wesley." *The New Encyclopedia of the American West.* Ed. Howard R. Lamar. New Haven, CT: Yale U Pr, 1998. 906–908.

693. Luton, Larry S. "Powell, John Wesley." *Encyclopedia of Environmental Issues.* Vol. 3. Ed. Craig W. Allin. Pasadena, CA: Salem Pr, 2000. 587–588.

694. Merrill, George P. "Powell, John Wesley." *Dictionary of American Biography.* Vol. 15. New York: Scribner's, 1935. 146–148.

695. Nelson, Clifford M. "Powell, John Wesley." *Encyclopedia of Earth Sciences.* Ed. E. Julius Dasch. New York: Simon & Schuster Macmillan, 1996. 887–889.

696. "Powell, John Wesley." *National Cyclopaedia of American Biography.* Vol. 3. New York: James T. White & Co., 1893. 340–341.

697. Stegner, Wallace Earle. "Powell, John Wesley." *Dictionary of Scientific Biography*. Vol. 11. Ed. Charles Couston Gillespie. New York: Charles Scribner's Sons, 1975. 118–120.

B. WESTERN EXPEDITIONS AND THE POWELL SURVEY, 1866–1878

Sections B–D are subdivided in two sections. The first (Contemporary Works) is for items published during Powell's life; the second (Post-19th-Century Works) is for material published after his death in 1902.

1. Contemporary Works

698. [Bishop, Will C.]. "Once More the Grand Canyon." *The Trail* 8.7 (1920): 16–19.

> Editor Bishop comments on Bill Bass's pamphlet, *Adventures in the Canyons of the Colorado* (**#87**), with lengthy extracts from Bill Hawkins' version of the incident at Separation Rapid in 1869 (**#122**).

699."The Cañons of the Colorado." *Baptist Home Mission Monthly* 4 (1882): 238–242.

> Describes the geography of the Colorado River country with quotations from *Exploration*.

700."The Canons [sic] of the Colorado." *Once a Week* [London] (1870): 184–190.

> The article quote remarks of explorers and surveyors, including Newberry, Bell, Ives, White, and Powell, on the Grand Canyon system.

701."The Colorado Expedition." *Annual of Scientific Discovery 1870*. Ed. John Trowbridge. Boston: Gould and Lincoln, 1870. 265.

> One of the first reports on the completed 1869 expedition.

702."Colorado River Canyons." *Mining and Scientific Press* 39 (1879): 193.

> Illustration of the "Mu-av" canyon with descriptive narrative from *Exploration*.

703. Daly, Charles Patrick. "Lands of the Arid Region of the West." *Journal of the American Geographical Society of New York* 10 (1878): 47–48.

> A summary of "the very interesting and important" Report on the Lands in "Geographical Work of the World in 1878 & 1879."

704. ——. “Major J. W. Powell’s Survey of the Rocky Mountain Region.” *Journal of the American Geographical Society of New York* 10 (1878): 30–31.

Summary of A. H. Thompson’s work in the field in “Geographical Work of the World in 1877.”

705.——. “Prof. J. W. Powell’s Exploration of the Colorado.” *Journal of the American Geographical Society of New York* 7 (1875): 46–48.

Report on the work of Powell Survey activities in 1877 in “The Geographical Work of the World for 1875.”

706.——. “Prof. Powell’s Exploration of the Colorado River.” *Journal of the American Geographical Society of New York* 4 (1873): 79.

Report on Powell’s continued survey of the Colorado River in “The Geographical Work of the World in 1872.”

707.——. “Professor Powell’s Expedition.” *Journal of the American Geographical Society* 8 (1876): 58–59.

Daly describes the work of members of the Powell expedition surveying the Eastern Utah region under the direction of A. H. Thompson. In “The Geographical Work of the World in 1876.”

708.——. “Professor Powell’s Explorations.” *Journal of the American Geographical Society of New York* 6 (1874): 65.

Summary of the 1872 report on the survey of the Colorado in “The Geographical Work of the World in 1874.”

709.“Editorial Notes.” *Illinois Wesleyan University Alumni Journal* 1 (1872): 167.

Report on Powell’s attendance at commencement and his farewell to friends. Also an announcement of H. C. DeMotte’s journey to Utah to join the Powell Survey.

710.“Exploration of the Colorado River of the West and Its Tributaries.” *Geological Magazine* [London] ns (Decade 2) 3 (1876): 365–370.

Positive review of Powell’s 1875 report. Discusses Powell’s geological observations in some detail.

711."Exploration of the Colorado River of the West and Its Tributaries." *American Journal of Science and Arts* 10 (1876): 303–304.

The reviewer states the work is largely a treatise on denudation and also gives much information on the Indians.

712."Geological Survey: The Grand Cañon Group." *Science* 1 (1883): 183–184.

Report on fieldwork by Powell, G. K. Gilbert, and Charles Walcott regarding the age of unconformable rocks in the Grand Canyon.

713. Godkin, E. L. "Our Unavailable Public Lands." *Nation* 26 (1878): 288–289.

The *Nation*'s editor supports Powell's *Report on the Arid Lands*. He maintains Powell's position that public lands should be classified and that divisional surveys should conform to topography and water sources.

714.——. "The Proposed Reform in Our Land and Scientific Surveys." *Nation* 28 (1879): 27–29.

Godkin here supports Powell's position on the disposal and management of arid lands, citing *Surveys of the Territories* (**#725**).

715.——. "Who Shall Direct Our National Surveys." *Nation* 18 (1874): 361–362.

This article expresses the views of the editor in support of Powell's position that civil scientists, not military engineers, should undertake the scientific work of the nation. This article sparked a series of remarks and letters in subsequent issues.

716. Hayden, Ferdinand V. *The Great West: Its Attractions and Resources.* Bloomington, IL: Charles R. Brodix, 1880.

A guidebook for travelers and practical compendium for homesteaders, as well as a history of Western exploration, and description of the wonders of Yellowstone. Hayden refers to Powell's explorations and quotes passages from *Exploration* (pp. 62–66).

717. Ingalls, George W. *Pi-Ute Indians. Letter From the Acting Secretary of the Interior, Relative to the Condition of the Pi-Ute Indians.* 42nd Cong., 3rd Sess., H. Ex. Doc. 66 Serial 1565. Washington: GPO, 1873.

Contains a letter in which Ingalls reports Powell's assistance as interpreter and skillful communicator with the Utah Paiute Indians.

718."Newspaper Reports of the Expedition's End." *UHQ* 15 (1947): 140–148.

Reprints of articles reporting on Powell's expedition in September 1869, from the *Deseret Evening News*, *Utah Daily Reporter*, *Chicago Tribune*, and *Cheyenne Leader*.

719."Powell's *Exploration of the Colorado River*." *American Naturalist* 10 (1876): 736–739.

Review of the second part of *Exploration* on structural geology; with illustrations.

720.["Powell's Return From the Grand Canyon"]. *Nature* 7 (1873): 290.

Brief summary of Powell's work in the summer and fall of 1872.

721. Rideing, William Henry. "The Canyons of the Colorado." *Potter's American Monthly* 19 (1882): 217–228.

Abridged version of the 1880 article (**#722**).

722.——. "A Chapter of American Exploration." *Lippincott's Magazine* 26 (1880): 393–410.

Rideing recounts the 1869 Colorado River expedition, based on the narrative from Powell's *Exploration*, and contrasts Western and Alpine exploration. See **#721** and **#723**.

723. ——. "A Chapter of American Exploration." *Here and There in Our Own Country*. Philadelphia: J. B. Lippincott, 1885. 116–133.

Reprint of **#722**.

724. Stanton, Robert Brewster. "The Denver, Colorado Canon [sic] & Pacific Railroad Project." *Engineering News and American Railway Journal* 22 (1889): 269–272.

In 1889, Robert Stanton led a Colorado River expedition through the Grand Canyon to survey for a railroad, during which the company's president, Frank Brown, drowned. Stanton blames Powell for Brown's death because Powell did not acknowledge that he had used a life preserver during the 1869 expedition. Stanton also disputes many of the details about the river in Powell's report, concluding that a railway could be constructed through the Canyon. See **#163** and **#164**.

725.*Surveys of the Territories. Letter From the Acting President of the National Academy of Sciences Transmitting A Report on the Surveys of the Territories.*

45th Cong., 3rd Sess., H. Misc. Doc. 5 Serial 1861. Washington: National Academy of Sciences, 1878.

Congress had ordered the NAS to look into duplication of the four Western surveys. The report contains Powell's *Report on the Methods of Surveying the Public Domain* (**#261**) and accompanying letters of his protagonists.

726. Whitney, James Dwight. "Geographical and Geological Surveys." *North American Review* 121 (1875): 37–85, 270–314.

Whitney assesses the numerous federal and state surveys, including Powell's, and decries the duplicate work of the Western surveys (pp. 79–85).

2. Post-19th-Century Works

727. Anderson, Martin J. "First Through the Canyon: Powell's Lucky Voyage in 1869." *Journal of Arizona History* 20 (1979): 391–408.

The author maintains it was "sheer luck" rather than careful planning that got Powell through the 1869 expedition. Highlights some of the losses and ordeals of the trip.

728.——. "John Wesley Powell's *Explorations* [sic] *of the Colorado River* ... Fact, Fiction, or Fantasy?" *Journal of Arizona History* 24 (1983): 363–380.

Argues that Powell manipulated the facts of his 1869 expedition and deliberately misleads his readers by presenting *Exploration* as one diary, rather than a refabrication of two "adventure" stories. Anderson draws heavily on Robert Stanton for evidence against Powell's character.

729. Bartlett, Richard Adams. *Great Surveys of the American West.* Norman: U of Oklahoma Pr, 1962.

A classic study of the accomplishments of the scientists, journalists, photographers, and others who constituted the geographical and geological surveys of the American West led by Wheeler, Hayden, Powell, and King. Bartlett includes the college field trips to Colorado in his overview of Powell's work as a government surveyor. Powell is given full credit for creating the USGS (pp. 219–329). Very good bibliography.

730. Belshaw, Michael. "The Dunn-Howland Killings: A Reconstruction." *Journal of Arizona History* 20 (1979): 409–422.

Belshaw posits that Shivwits murdered Dunn and Howlands on the last day of August or first day of September 1869, at the base of Mt. Dellenbaugh. He speculates that the motive was robbery. See **#741** for response by Dobyns and Euler.

731. Brooks, Juanita. "Lee's Ferry at Lonely Dell." *UHQ* 25 (1957): 283–295.

History of Lee's Ferry, where the Paria River meets the Colorado and members of the Powell party were frequent visitors.

732. Childs, Elizabeth C. "Time's Profile: John Wesley Powell, Art, and Geology at the Grand Canyon." *American Art* 10.1 (1996): 7–35.

Childs focuses on the work of the photographers and artists of the second Colorado River expedition. She provides an overview of the techniques and products (stereopticons) of Powell photographers Beaman, Fennemore, Clem Powell, and Hillers, as well as Powell's impact on Thomas Moran and Moran's on Powell. Twenty illustrations.

733. Corle, Edwin. *Listen, Bright Angel.* New York: Duell, Sloan and Pearce, 1946.

An entertaining and informative history of the geology and people of the Grand Canyon, including Powell's experiences there (pp. 113–127). Corle notes the impact that the canyon had on Powell.

734. Crampton, C. Gregory. *Land of Living Rock; the Grand Canyon and the High Plateaus: Arizona, Utah, Nevada*. New York: Alfred A. Knopf, 1972.

This fully illustrated "historical portrait" presents the geology, history, and ecology of the Grand Canyon and the high plateau country. Chapter 7 is devoted to Powell's 1869 expedition and the status of the Native American population at that time (pp. 113–133).

735. Darrah, William Culp. "Geological Notes by Major Powell." *UHQ* 15 (1947): 132–133.

Background on Powell's small notebook from 1869, containing geological notes and sections, as well as various tables of astronomical data. See **#9**, **#32**, and **#31**.

736.——. "John Wesley Powell, His Western Explorations." *Geotimes* 14 (1969): 13–15.

Overview of Powell expeditions and surveys, 1867–1877.

737.——. "Major John Wesley Powell Prepares for a Second Expedition." *UHQ* 15 (1947): 149–153.

This article details Powell's preliminary work for the second Colorado River expedition and his interaction with the Mormons and Indians of Utah.

738.——. "The Powell Colorado River Expedition of 1869." *UHQ* 15 (1947): 9–18.

In addition to providing historical background for the 1869 expedition, Darrah reviews the source materials and documents of the river party used in this volume of *UHQ* (see **#102**). Includes comments on Robert Stanton's "distortions" and the importance of Bradley's journal.

739. Dellenbaugh, Frederick Samuel. *Breaking the Wilderness: The Story of the Conquest of the Far West, From the Wanderings of Cabeza De Vaca, to the First Descent of the Colorado by Powell, and the Completion of the Union Pacific Railway, With Particular Account of the Exploits of Trappers and Traders.* New York: G. P. Putnam's Sons, 1905.

A review of historical events leading to the settlement of the American West. Chapter 16 includes an account of the 1869 expedition. Illustrations include drawings by Dellenbaugh and photographs by Hillers. Dedicated to A. H. Thompson.

740.——. "Naming the Grand Canyon." *Science* ns 77 (1933): 349–350.

Dellenbaugh claims Powell was the first to explore and name the Grand Canyon.

741. Dobyns, Henry F., and Robert C. Euler. "The Dunn-Howland Killings: Additional Insights." *Journal of Arizona History* 20 (1980): 87–95.

Direct response to Belshaw's article (**#730**). The authors disagree with his dismissal of testimony from Shivwit Paiutes regarding the murder of Dunn and the Howland brothers. Dobyns and Euler claim that the three were killed as a continuation of a war against "Euroamerican invaders" by Shivwits and Pine Springs Band Paiutes.

742. Dolnick, Edward. *Down the Great Unknown; John Wesley Powell's 1869 Journey of Discovery and Tragedy Through the Grand Canyon.* New York: HarperCollins, 2001.

Drawing on published journals (primarily those in *UHQ*), letters, and later accounts written by members of the 1869 expedition, Dolnick recreates a narrative, day-by-day account of the 1869 expedition. Published in London with minor changes in 2002.

743. *Exploring the American West, 1803–1879*. National Park Handbook 116. Washington: U.S. Dept. of the Interior, National Park Service, 1982.

Story of how the West was surveyed, mapped, and explored, with an overview of the pictorial record. Richard Bartlett describes "The Lure of the West" (pp. 7–15); William Goetzman discusses the 1869 expedition and the Powell Survey in "Explorer, Mountain Man, and Scientist" (pp. 19–95).

744. Ferris, Robert G., ed. *The American West: An Appraisal*. Santa Fe: Museum of New Mexico Pr, 1963.

Twenty papers from the second Conference on the History of Western America. Contains an essay by Bartlett, "John Wesley Powell and the Great Surveys" (pp. 48–57). See **#729**.

745. Frame, Paul N. "The Great Surveyors: Hayden, King, Powell and Wheeler." *Mountain Plains Library Quarterly* 8.4 (1964): 19, 20–22.

An overview of the publications of the predecessor surveys.

746. Ghiglieri, Michael P., et al. *First Through Grand Canyon; the Secret Journals and Letters of the 1869 Crew Who Explored the Green and Colorado Rivers*. Flagstaff, AZ: Puma Pr, 2003.

Ghiglieri maintains that none of those who have attempted to describe the 1869 expedition "managed to get it right or even complete." He agrees with Robert Stanton that Powell's 1875 report and *Canyons of the Colorado* slight the crew and are historically inaccurate. He finds hundreds of errors in Darrah's and others' transcriptions of the journals and omissions from Powell's letters to the *Chicago Tribune* in *UHQ* 15 (see **#568**). Ghiglieri's narrative of the 1869 voyage is compiled from his own transcriptions of letters and diaries written by Powell and members of his crew while on the river, including George Bradley, Andrew Hall, William Hawkins, O. G. Howland, Walter H. Powell, and Jack Sumner. Includes biographies and an appended transcription by Richard Quartaroli of Powell's astronomical journal (**#32**). See **#9**, **#38**, **#57**, **#58**, **#97**, **#119**, **#126**, **#138**, and **#140**.

747. Ghiglieri, Michael P., and Thomas M. Myers. *Over the Edge: Death in the Grand Canyon*. Flagstaff, AZ: Puma Pr, 2001.

Within the context of investigating murders that took place in the Grand Canyon, Ghiglieri and Myers evaluate Powell as a leader of men (pp. 315–333). They contend that his failure resulted in the murder of three brave men. The authors rely on the accounts of Sumner and Hawkins (**#123**, **#137**) and Wesley Larsen (**#752**).

748. Goetzmann, William H. *Exploration and Empire; the Explorer and the Scientist in the Winning of the American West*. New York: Alfred A. Knopf, 1966.

Beginning with the travels of Lewis and Clark, Goetzmann presents a well-documented, thorough history of 19th-century exploration of the American West, much of which was official exploration or institutionally oriented. Goetzmann assesses Powell as a self-made scientist who, like other late-19th-century politicians and intellectuals, was a reformer who worked for the public good (pp. 530–576). The final chapter is an epilogue that closes the age of discovery in the West with the formation of the USGS. Includes excellent bibliographic notes on sources. Along with Bartlett (**#729**), Smith (**#864**, **#865**), and Dupree (**#850**), this important work lifted Powell from historical obscurity.

749. Gregory, Herbert E. "Scientific Explorations in Southern Utah." American Journal of Science 243 (1945): 527–549.

Overview of activities of exploring parties in Southern Utah, with emphasis on the work between 1842 (Fremont) and 1880 (the four Western surveys).

750. Ivins, Anthony W. "Traveling Over Forgotten Trails." *Improvement Era* 27 (1924): 1017–1025. [Rpt. "A Mystery of the Grand Canyon Solved." *Pioneer Stories*. By Preston Nibley. Salt Lake City: Deseret News Pr, 1940. 199–210.]

Ivins claims to have found the site where the Howlands and Dunn were murdered by an Indian.

751."John Wesley Powell." *Great Explorers of the West*. National Survey of Historic Sites and Buildings; Theme XV, Westward Expansion and Extension of the National Boundaries, 1830–1898. [Washington?]: National Park Service, 1960. 40–45.

"The purpose of this study is to assemble data on historic sites believed to be of exceptional value in commemorating or illustrating the history of western exploration." The chapter on Powell (pp. 40–45) gives a brief account of 1869 expedition.

752. Larsen, Wesley P. "The 'Letter' or Were the Powell Men Really Killed by Indians?" *Canyon Legacy* 17 (1993): 12–19.

Larsen introduces and reprints a letter from Mormon William Leany to John Steele regarding the murder of three men near Toquerville, Utah, in 1869. Larsen postulates that the men are Dunn and the Howland brothers, though the evidence is admittedly circumstantial.

753. Lavender, David. *Colorado River Country*. New York: Dutton, 1982.

A history of Euro-American exploration of the Colorado River Plateau, including Powell's expeditions.

754.——. "First Conquest of the Colorado; the Powell Expeditions." *National Parks* 60.5–60.6 (1986): 16–21.

Account of both Powell Colorado River expeditions. Powell didn't write an official report of 1869 expedition until 1874 under pressure from James Garfield. He used events and place names related to the 1871–1872 runs to present a unified narrative.

755. MacKaye, Benton. "Twin American Conquests: Pacific Railroad and Grand Canyon." *Cosmos Club Bulletin* 22.3 (1969): 4–7.

The author presents facts regarding the completion of the continental railroad and the beginning of Powell's historic voyage in May 1869. He recalls seeing a presentation by Powell in 1891 at the National Geographic Society.

756."Major Powell and the Colorado." *Nevada Highways and Parks* 20.2 (1960): 18–24.

Excerpts from *Exploration* and highlights of the feature film, *Ten Who Dared* (**#1010**); illustrated with still photographs.

757. Marston, Otis R. "Early Travel on the Green and Colorado Rivers." *The Smoke Signal* (Tucson Corral of the Westerners) 20 (1969): 231–236.

A brief history of river crossing and journeys on the Green and Colorado Rivers through 1869. Marston examines Powell's expedition from a boatman's point of view.

758.——. "Separation Marks: Notes on 'the Worst Rapid' in the Grand Canyon." *Journal of Arizona History* 17 (1976): 1–20.

Separation Rapid is where the Howlands and Dunn left the Powell expedition in 1869. There are many controversies surrounding the incident, including whether they left because Powell forced them to or because they were cowards. "Dock" Marston traces river runners' encounters with the rapids, from Powell through 1934.

759.——. "With Powell on the Colorado." *Westerners Brand Book 2: The San Diego Corral.* 1971. 65–76.

Marston follows Powell's 1869 river trip, criticizing his skills as a riverman and questioning the accuracy of his writings. Powell earned his reputation in science and for his political activities, but as a boatman he was just lucky, Marston contends.

760. Murphy, Dan, and Gary Ladd. *John Wesley Powell: Voyage of Discovery.* Story Behind the Scenery. Las Vegas, NV: KC Publications, 1991.

This version of the 1869 Colorado River expedition is illustrated with maps and full-color photographs by Gary Ladd.

761. Olsen, Robert W., Jr. "Pipe Spring, Arizona, and Thereabouts." *Journal of Arizona History* 6 (1965): 11–20.

The early history of Pipe Spring, Arizona, where the paths of Powell, Jacob Hamblin, Francis Bishop, and Brigham Young converged in 1870.

762. ——. "The Powell Survey Kanab Base Line." *UHQ* 37 (1969): 261–268.

This article is concerned with technical aspects of the survey's work in 1871–1872. The first base line measured in the area was bounded by Colorado on the east and the Sierra Nevada Mountains on the west. From this line, the men systematically explored and surveyed the region.

763. Pickering, James H. *This Blue Hollow: Estes Park, the Early Years, 1859–1915.* Niwot: U Pr of Colorado, 1999.

Chapter 7 includes an account of the Powell party's ascent of Long's Peak in 1868 and the experience of later climbers who found the tower of stones erected by Powell on the summit. The notes left by Powell in the tower were removed in 1935 and given to the Colorado Historical Society.

764. Pike, Donald G. "Four Surveyors Challenge the Rocky Mountain West: Fighting Bureaucracy and Indians in a Wild Land." *American West* 9.3 (1972): 4–13.

An overview of the men who led the four Western surveys—Powell, Hayden, King, and Wheeler.

765. Powell, Lawrence Clark. "The Man Who Ran the River." *Westways* 46.6 (1972): 37–39, 50–51.

A bibliographic essay on *Exploration*, with background information on the life of Powell.

766. Rabbit, Mary C. *John Wesley Powell's Exploration of the Colorado River.* USGS: INF-74-19. Washington: GPO, 1974. [Rpt. 1976, 1978, 1981, and 1995.]

This booklet on the 1869 expedition includes passages from *Exploration* and photographs from the second Colorado River expedition.

767.——. "Famous Trip Boosts Powell's Career." *Reclamation Era* 55.1 (1969): 6–10.

A review of Powell's career in the federal government, particularly his work with the Powell Survey and the USGS. Powell's ideas on the use of water and arid lands were vindicated by the establishment of the Bureau of Reclamation in 1902.

768.——. "Utah 1869: Explorations and Expeditions." *Proceedings of the Utah Academy of Sciences, Arts and Letters* 46, pt. 2 (1969): 27–35.

A historical snapshot of Utah in 1869, when three of the Western surveys (those of Powell, King, and Wheeler) were within less than 100 miles of each other and the first transcontinental railroad was completed.

769. Roscoe, Gerald, and David Larkin. *Westward: The Epic Crossing of the American Landscape.* New York: Monacelli Pr, 1995.

The authors devote a chapter to Powell's 1869 exploration of the Grand Canyon, with excerpts from *Exploration.*

770. Rusho, W. L. "Francis Bishop's 1871 River Maps." *UHQ* 37 (1969): 207–213.

The six remaining map sheets from the 1871–1872 Powell expedition are reproduced with background information and commentary by Rusho. He believes that these sheets, preserved in the archives of the Utah State Historical Society, are Bishop's personal copies (see also **#1000**).

771.——. *Powell's Canyon Voyage.* Wild & Woolly West Books 11. Palmer Lake, CO: Filter Pr, 1969.

This small book presents the events of the 1869 Colorado River expedition, illustrated with engravings of the Grand Canyon reproduced from various reports by Powell, Dutton, and Stanton.

772. Scher, Zeke. "On the Trail of John Wesley Powell." *Empire Magazine* (*Denver Post*) 77.290 (May 18, 1969): 8–15 and 77.297 (May 25, 1969) 14–23.

Two-part series in the *Denver Post*'s Sunday magazine on Powell's travels in Colorado in 1867–1868 and his 1869 voyage down the Colorado River, with biographical information.

773. Smith, Melvin. "Before Powell: Exploration of the Colorado River." *UHQ* 55 (1987): 104–119.

An overview of the human history of the Colorado River before 1869. Smith looks at the Indians, Spaniards, miners, merchants, Mormons, government surveyors, and others who explored the region and recorded its history.

774. Stanton, Robert Brewster, James McGowan Chalfant, and Julius Stone. *Colorado River Controversies*. New York: Dodd Mead & Company, 1932.

This is the second half of Stanton's manuscript on the Colorado River, edited by James M. Chalfant with a foreword by Julius F. Stone. It contains a historical overview of river expeditions and voyages as well as Stanton's theory of the geological history of the country. The portions published here relate to two of the most controversial events in the Colorado's history. The first section, "James White's Raft Journey of 1867," is Stanton's investigation of White's claim to be the first to navigate the river. In "The Affair at Separation Rapid," Stanton presents two documents from William Hawkins (**#123**) and John Sumner (**#137**), written after the events of 1869, defending Dunn and the Howlands against charges of cowardice. Chalfant and Stanton are highly critical of inaccuracies regarding the Colorado River and the Grand Canyon in Powell's 1875 narrative.

The second half of Stanton's manuscript was later published as *Down the Colorado* (**#164**). See **#8** and **#775**.

775. Stanton, Robert Brewster, et al. *Colorado River Controversies*. 1932. Boulder City, NV: Westwater Bks, 1982.

This unabridged reproduction of the 1932 publication is appended with essays by Otis Marston (pp. 233–251) and Martin Anderson (pp. 253–261). Marston comments on White's controversial claim to being the first to descend the Grand Canyon in light of evidence made available since the 1932 publication. He disagrees with Stanton's view that White's claims were impossible. Anderson comments on Powell, correcting several statements by Stanton, Stone, and James Chalfant (**#774**).

776. Stegner, Wallace Earle. "Jack Sumner and John Wesley Powell." *Colorado Magazine* 26 (1949): 61–69.

Stegner examines a letter from Jack Sumner printed in the *Denver Post* in October 1902, reproduced in this article. The letter accuses Powell of poor leadership and not meeting his financial obligations to the men of the 1869 expedition. Stegner asserts Stanton did not have all the evidence and that his correspondence with Sumner and Hawkins "encouraged criticism of Powell by leading questions."

777. ——. "Powell and the Names on the Plateau." *Western Humanities Review* 7 (1953): 105–110.

In this excerpt from *Beyond the Hundredth Meridian* (**#671**), Stegner tells the story of the names given by Powell and Dutton to geographical features of the Colorado River region.

778. Steward, Julian Haynes, and John K. Hillers. *Notes on Hillers' Photographs of the Paiute and Ute Indians Taken on the Powell Expedition of 1873.* Smithsonian Miscellaneous Collections 98, no. 18. Washington: Smithsonian Institution, 1939.

When Steward compiled Hillers' photographs for this publication, none of Powell's ethnographic work had been published as a body. These photographs, with Steward's careful documentation, provide important ethnological and anthropological data on the Utes and Paiutes before their native culture was eradicated by contact with settlers and Indian agents.

779. Tikalsky, Frank D. "Historical Controversy, Science, and John Wesley Powell." *Journal of Arizona History* 23 (1982): 407–422.

Tikalsky, a psychology professor, analyzes Powell's leadership style and the events at Separation Rapid in terms of behavioral norms exhibited by individuals participating in long, stressful expeditions. He concludes that the conflicting views of Powell held by members of the 1869 expedition are not untypical.

780. Toll, David. "The Powell Exploration of the Green and Colorado Rivers 1869–1969." *Arizona Highways* 45.3 (1969): 8–11, 37–43.

Toll narrates his journey following the path of Powell's voyage, by car and foot from Green River to Marble Canyon and by boat to Diamond Creek. Interspersed with text from *Exploration*.

781. Truettner, William, and Nancy K. Anderson. *The West As America: Reinterpreting Images of the Frontier, 1820–1920*. Washington: Smithsonian Institution Pr, 1991.

This collection of essays examines how artists used their talents on various explorations and scientific expeditions in the West, including Thomas Moran and others who worked with the Powell Survey (pp. 248–259).

782. Watkins, T. H., et al. *The Grand Colorado: The Story of a River and Its Canyons*. Palo Alto, CA: American West, 1969.

Foreword by Wallace Stegner. This oversize book on the human history of the Colorado River region includes a chapter on the first Powell expedition (pp. 89–141). Introductory material contains Powell's biography and excerpts from *Exploration*, with commentary by Stegner from *Beyond the Hundredth Meridian* (**#671**).

783. Watson, Elmo Scott. *The Illinois Wesleyan Story 1850–1950*. Bloomington: Illinois Wesleyan U Pr, 1950.

Watson describes Powell's years at IWU, where he served as professor of natural sciences 1865–1868. Powell designed the seal of the university and led students and amateur scientists on expeditions to Colorado in 1867 and 1868.

784.——. "John W. Powell's Colorado Expedition of 1867." *Colorado Magazine* 27 (1950): 303–311.

Watson describes the 1867 expedition to the Rocky Mountains. He reprints text from two students, one a manuscript by Milton Titterington and letters to the Bloomington (Illinois) *Daily Pantagraph* by Joseph Hartzell. Similar to portions of *Professor Goes West* (**#144**).

785. Webb, Roy. *Call of the Colorado*. Moscow: U of Idaho Pr, 1994.

This illustrated history of the scientists, surveyors, outfitters, and others who documented and explored the Colorado River begins with Powell's 1869 voyage.

786.——. "John Wesley Powell, the Other Expedition." *Utah Outdoors* 14 (September 2000): 30–33, 62–63.

The story of the second Colorado River expedition.

787. Weisheit, John. "Place Names Along the Rivers in Utah's Canyonlands." *Canyon Legacy* 18 (1993): 24–31.

Powell was the significant name-maker of places along the Colorado and Green Rivers. Weisheit correlates contemporary places names with those given by Powell and his companions.

788. Wharton, James George. *In and Around the Grand Canyon: The Grand Canyon of the Colorado River in Arizona.* Boston: Little, Brown, 1900.

Wharton dedicated this tourist guide to Powell. He devotes a chapter to Powell's Colorado River expeditions.

789. Wheat, Carl I. *Mapping the Transmississippi West 1540–1861; Vol. 5, pt. 2: From the Civil War to the Geological Survey.* San Francisco: Institute of Historical Cartography, 1963.

This beautiful six-volume set is an exceptional cartographic bibliography. V. 5 provides historical context for Powell's early explorations and the formation of the USGS. Wheat discusses the naming of the Grand Canyon by William Palmer before the 1869 expedition (pp. 254–256) and the maps of the four Western surveys (pp. 361–369).

790. Wilkins, Thurman. "Major Powell & Thomas Moran in Canyon Country." *Montana* 19.4 (1969): 16–31.

Powell invited Moran to join him in 1872. Moran had other plans, but he took up the offer in 1873. This article describes Moran's travels in Utah with journalist Justin Colburn from the *New York Times* and the art that followed.

791. Worster, Donald. "The Second Colorado River Expedition: John Wesley Powell, Mormonism and the Environment." *Surveying the Record: North American Scientific Exploration to 1930.* Ed. Edward Carlos Carter. Memoirs of the American Philosophical Society 231. Philadelphia: American Philosophical Society, 1999. 317–328.

During the second Colorado River expedition, Powell relied on assistance and guidance from Mormon settlers in Utah. Worster examines the lasting influence of his experiences among the Mormons on Powell's scientific research and the development of his unconventional policies for settlement and management of arid lands. See **#917**.

C. DIRECTOR OF THE BUREAU OF AMERICAN ETHNOLOGY, 1879–1902

1. Contemporary Works

792. Gore, J. Howard. "Anthropology at Washington." *Popular Science Monthly* 35 (1888): 786–795.

Describes the history and work of the governmental institutions engaged in collecting information about the Indians of North America: the Smithsonian Institution National Museum, the BAE, and the Army Medical Museum. Includes a bibliography of the principal publications of the BAE, 1878–1887.

793. Hartland, E. Sidney, and N. W. Thomas. "An American View of Totemism: [Notes]." *Man* [London] 2 (1902): 113–118.

Two separate responses to Powell's article (see **#544**) provide insight into some of the differences between American and European anthropology.

794."Major Powell Condemned by the Press: Criticisms by Prominent Scientists." *The Naturalist* (Kansas City) 4.9 (1890): 1.

A report on letters received in support of Warren Watson's attack on Powell in the February 1890 issue (see **#796**).

795."*Twentieth Annual Report of the Bureau of American Ethnology to the Secretary of the Smithsonian Institution, 1898–'99.*" *Athenaeum* No. 3999 (June 18, 1904): 788.

Review of the BAE report, including a summary of Powell's appended papers. The reviewer notes, "His observations are enlivened by shrewd and apposite deductions from the experience of his life's work ... but the desire to find a pentalogy in everything has led to somewhat fanciful distinctions."

796. Watson, Warren. "Those Elephant Pipes Again." *The Naturalist* (Kansas City) 4.7 (1890): 1–2.

An attack on Powell for remarks made in his article, "Prehistoric Man in America" (**#409**), regarding a pair tusks held by the Davenport Academy of Natural Sciences. See **#794** for subsequent remarks.

797. Wickersham, James. *Major Powell's Inquiry "Whence Came the American Indians?" An Answer: A Study in Comparative Ethnology*. Tacoma, WA: Allen & Lamborn Printing Co, 1899.

In response to Powell's article in *Forum* (**#523**), Wickersham disagrees with Powell's answer to the question posed. He cites ethnological evidence supporting his view that the origin of the American Indian's origin is Mongolian tribes, who recently occupied the New World.

2. Post-19th-Century Works

798. Boas, Franz. *Handbook of American Indian Languages*. BAE Bulletin 40. Washington: GPO, 1911.

From the preface: "The *Handbook* ... had its inception in an attempt to prepare a revised edition of the 'Introduction to the Study of Indian Languages.'" Boas edited the first two volumes in this series of linguistic sketches, which continues to be of value to linguists and anthropologists. (Part 3 was later published in 1938 and a section of Part 4 in 1941). Boas's introduction was later reprinted with Powell's "Indian Linguistic Families of America North of Mexico" (**#577**, **#590**).

799. Darnell, Regna. "Powell Classification of American Indian Languages." *Papers in Linguistics* 4 (1971): 71–110.

This paper is a revision of a section of Darnell's dissertation (**#983**). Darnell examines Powell's Indian language classification, with attention to the social context of the BAE, the evaluation of linguistic data, and the state of knowledge of Indian languages in 1891. Following the bibliography is a table that summarizes key information about each language group.

800.——. "Revision of the Powell Classification." *Papers in Linguistics* 4 (1971): 233–257.

Revision of a section from Darnell's dissertation (**#983**). Darnell points out that Powell's aim in classifying Indian languages was not linguistic but pragmatic: classification of tribes would permit better administration of Indian settlement. Powell's classification was subsequently revised by Boas, A. L. Kroeber, and Edward Sapir.

801. De Laguna, Frederica, ed. *American Anthropology, 1888–1920; Papers From the* American Anthropologist. Lincoln: U Nebraska Pr, 1960.

Contains 55 selections from the *American Anthropologist*, with comments by De Laguna preceding each section. An introductory essay by Alfred Irving Hallowell presents a comprehensive overview of the development of American anthropology, including Powell's role. Reprints of articles by and about Powell include his comments on "The Nomenclature and Teaching of Anthropology" (**#435**); "Stone Art in America" (**#485**); and "Obituary of John Wesley Powell, March 13, 1834–September 23, 1902" (**#639**). Paperback reprint 2002.

802. Douglas, Frederic H. *Indian Linguistic Stocks or Families; Powell's Classification and Modern Changes; Tribes, Locations and Population.* Vol. 2. Leaflets 51–52. Denver, CO: Denver Art Museum, Dept. of Indian Art, 1933.

This revision of the original 58 language stocks classified in "Indian Linguistic Families of America North of Mexico" (**#422**) updates census information and other changes in population from several sources, including *Handbook of American Indians North of Mexico* (**#812**) and the 1910 census of American Indians.

803. Fowler, Don D., and Catherine S. Fowler. "John Wesley Powell, Anthropologist." *UHQ* 37 (1969): 152–172.

Overview of Powell's achievements as an ethnologist, anthropologist, and linguist. The authors' assessment is that his ethnographic and linguistic fieldwork and his work in linguistic classification are exemplary. They also argue that he was also instrumental in the transition of American anthropology in the 19th century from "its antiquarian eclecticism" to its status as a social science.

804. ——. "Notes on the History of the Southern Paiutes and Western Shoshonis." *UHQ* 39 (1971): 95–113.

This article focuses on events in Utah and Arizona during the period of Anglo exploration and occupation that forced changes in native societies and cultures. Included are reproductions of two letters written in 1880 by Jacob Hamblin to Powell detailing the plight of the Southern Paiutes near Kanab, as well as Powell's letter in response. Powell's views on the Indians of the Great Basin are conveyed in *Report of Special Commissioners* (**#236**).

805. Fowler, Don D., and John F. Matley. *Material Culture of the Numa: The John Wesley Powell Collection, 1867–1880.* Smithsonian Contributions to Anthropology 26. Washington: Smithsonian Institution Press, 1979.

This important work is a comprehensive study of the John Wesley Powell Collection in the Smithsonian Institution National Museum of Natural History. Photographs and illustrations present artifacts and photographs of the Numic-speaking Indian groups (Ute and Paiute). It is the only significant publication to date on this topic.

806. Fowler, Don D., Robert C. Euler, and Catherine S. Fowler. *John Wesley Powell and the Anthropology of the Canyon Country.* Geological Survey Professional Paper 670. Washington: U.S. Dept. of the Interior, Geological Survey, 1969. [Rpt. Grand Canyon Natural History Association, 1977 and 1981.]

A study of Powell's anthropological fieldwork and the archeology and ethnology of the Colorado River canyons, with extracts from Powell's manuscripts in the NAA (**#9**) on the origins and customs of the Indians of that area. An essential work.

807. Goddard, Ives. "The Classification of the Native Languages of North America." *Handbook of American Indians.* Vol. 17: Languages. Ed. William C. Sturtevant. Washington: Smithsonian Institution, 1996. 290–323.

In the section titled "Powell's Classification" (pp. 299–308), Goddard analyzes Powell's classification system in "Indian Linguistic Families of America North of Mexico." He compares it with other classifications and documents fieldworkers who compiled the data. Changes and revisions to the accompanying map, *Linguistic Stocks of American Indians North of Mexico*, are described and tracked. Heavily referenced article with a comprehensive bibliography.

808. Haas, Mary. "Grammar or Lexicon? The American Indian Side of the Question From Duponceau to Powell." *International Journal of American Linguistics* 35 (1969): 239–255.

A chronological review of the efforts throughout the 19th century to classify American Indian languages, concluding with Powell's "Indian Linguistic Families of America North of Mexico" (**#422**). Those who attempted a classification relied on lexicon as the primary criterion of differentiation. Powell based his model on Linnean classification.

809. Hinsley, Curtis Matthew, Jr. "Anthropology as Science and Politics: The Dilemmas of the Bureau of American Ethnology, 1879–1904." *The Uses of Anthropology*. Ed. Walter Goldschmidt. AAAS Special Publication 11. Washington: American Anthropological Association, 1979. 15–32.

> According to Hinsley, Powell was a reformist who believed that social action should be based on positive science. Under Powell's direction, American anthropology provided the service of "moral utility" in its approach to providing the federal government with information to formulate public policy regarding the native population.

810.——. *Savages and Scientists; The Smithsonian Institution and the Development of American Anthropology, 1846–1910*. Washington: Smithsonian Institution Pr, 1981.

> Republished with new title and preface. See **#811** for annotation.

811.——. *The Smithsonian and the American Indian; Making a Moral Anthropology in Victorian America*. Washington: Smithsonian Institution Pr, 1994.

> A thorough, well-documented intellectual history of American anthropology in the 19th century and an insightful analysis of Powell's place in it. According to Hinsley, the founding of the BAE under Powell's leadership changed the direction of American anthropology. As a moral philosopher, Powell looked to the past for clues to the future. He created a hierarchical organization in which one class of ethnologist undertook the preliminary work of collection and observation, while the higher class synthesized and hypothesized. Hinsley concludes that Powell and the Washington anthropologists of the late 19th century celebrated humanity in the face of an increasingly technological, corporate environment; at the same time, they unintentionally paved the way for modern anthropology in the next century. Originally published as *Savages and Scientists* (**#810**).

812. Hodge, Frederick Webb, ed. *Handbook of American Indians North of Mexico*. BAE Bulletin 30. Washington: GPO, 1907.

> The work for this handbook began as early as 1873 when Otis Mason began preparing a list of tribal names. Powell and BAE scholars continued the work, publishing *Linguistic Stocks of American Indians North of Mexico* (**#417**) in 1891. In 1902, work on the handbook resumed, its subject matter enlarged, and culminated in this two-volume publication edited by Hodge. Each linguistic stock is briefly described and supplemented with information on location, population, and the history of every associated confederacy, tribe, or tribal division.

813. Hymes, Dell H. "Kroeber, Powell, and Henshaw." *Anthropological Linguistics* 3.6 (1961): 15–16. [Rpt. *Anthropological Linguistics* 35 (1993): 51–53.]

> Hymes supports the view of anthropologist Alfred Kroeber that Henry Henshaw was the real author of Powell's linguistic classification system. See **#815**.

814. Judd, Neil M. *The Bureau of American Ethnology: A Partial History.* Norman: U of Oklahoma Pr, 1967.

> This compact but thorough history of the BAE contains a comprehensive list of publications through 1964 (pp. 78–134), when the bureau merged with the Division of Ethnology of the National Museum to form the Smithsonian Office of Anthropology. Also a useful source of information on BAE researchers.

815. Kroeber, Alfred L. "Powell and Henshaw: An Episode in the History of Ethnolinguistics." *Anthropological Linguistics* 2.4 (1960): 1–5. [Rpt. *Anthropological Linguistics* 35 (1993): 46–50.]

> Recalling his work in the BAE, Kroeber identifies the anthropologists and linguists responsible for Powell's classification of Indian languages. Powell was the organizer of the work and creator of the system, but staff linguists did the collection and made the comparisons and preliminary judgments. Powell needed a middleman—Henry Henshaw—to pull it all together. Kroeber also comments on the differences between Powell and his competitor in linguistics, Daniel Brinton. See **#813**.

816. Lacey, Michael J. "The World of the Bureaus: Government and the Positivist Project in the Late Nineteenth Century." *The State and Social Investigation in Britain and the United States.* Ed. Michael J. Lacey and Mary O. Furner. Washington and Cambridge: Woodrow Wilson Center Pr and Cambridge U Pr, 1993. 127–170.

> Powell was the most prominent spokesman for the intellectual community in Washington in the late 19th century. As head of BAE, he was concerned with problems of theory in accounting for social evolution. Powell and others practiced "new liberalism ... based on an evolutionary view of the role of the state as a moral agency in social development."

817. Lamb, Daniel S. "The Story of the Anthropological Society of Washington." *American Anthropologist* ns 8 (1906): 564–579.

A history of the ASW, its publications, and presidential addresses, including Powell's.

818. Longacre, William A. "Why Did the BAE Hire an Architect?" *Journal of the Southwest* 41(1999): 359–69.

Architect Victor Mindeleff was hired by Powell to explore the relationship between the architecture and the social organization of Southwestern American Indians.

819. Shaul, D. Leedom. "Linguistic Natural History: John Wesley Powell and the Classification of American Languages." *Journal of the Southwest* 41 (1999): 297–310.

This essay examines Powell's place in the history of American Indian anthropological linguistics. Powell's classification system used "the lexical inspection method." The linguistic work of Powell and his colleagues at the BAE laid the foundation for reconstructing the prehistory of the Southwest prehistory and continues to provide evidence for ethnic diversity.

820. Sturtevant, William C. "Authorship of the Powell Linguistic Classification." *International Journal of American Linguistics* 25 (1959): 196–199.

Sturtevant makes the case that although Powell claimed he alone was responsible for the classification of languages presented in "Indian Linguistic Families of America North of Mexico" (**#422**), the work was "a cooperative undertaking" involving many BAE employees and associates.

821. Woodbury, Richard B., and Nathalie F. S. Woodbury. "The Rise and Fall of the Bureau of American Ethnology." *Journal of the Southwest* 41 (1999): 283–296.

Woodbury traces the history of the BAE from its origin in the establishment of the Smithsonian Institution in 1846 through its merger with the Museum of Natural History in 1964. Powell's embrace of Lewis Morgan's theory of cultural evolution formed the BAE's intellectual foundation.

D. DIRECTOR OF THE U.S. GEOLOGICAL SURVEY AND THE IRRIGATION SURVEY, 1894–1902

1. Contemporary Works

822. Baker, Marcus. "The U.S. Geological Survey." *Geographical Journal* 6 (1895): 252–260.

In a tribute to Powell and the work of the USGS since its founding, Baker pays particular attention to mapping. He explains that only 12 of the atlas sheets of a geological map have been published, because it was first necessary for the survey to complete topographical maps.

823. Cope, Edward D., and Kingsley J. S. "Editorial." *American Naturalist* 24 (1890): 460–462.

As part owner of *The American Naturalist,* Cope had ample opportunity to attack Powell in print. Here he and Kingsley equate the motion of Powell to move the International Congress of Geologists from Philadelphia to Washington with "the destruction of the international character of the Congress."

824."Correspondents' Column." *The Manufacturer and Builder* 20 (1888): 287.

Question to and answer from the editor regarding the Sundry Civil Appropriation bill, which authorized funding for the Irrigation Survey.

825."Editorial Comment." *American Geologist* 12 (1893): 115–118.

The editor takes issue with Powell's article in *Popular Science*, "Are There Evidences of Man in the Glacial Gravels?" (**#441**). He is accused of inaccuracy and failure to answer criticisms posed in a previous article.

826. Frazer, Persifor. "An Unjust Attack." *American Geologist* 3 (1889): 65–72.

Frazer replies to articles by J. D. Dana and Powell on the American Committee of the International Congress of Geologists (see **#364**, **#352**). Contending that Dana and Powell are inconsistent in their charges against the committee, Frazer points out that Powell insists on freedom of thought on the one hand, yet objects to the various independent points of view of the committee's members.

827. Gannett, Henry. "The Topographic Work of the National Geological Survey." *American Geologist* 11 (1893): 65–67, 127–128.

In response to an editorial critical of the USGS's topographic mapping (see **#844**), chief geographer Gannett defends the importance and quality of the work.

828. Hancock, William A. "In Reply to Powell: The Duty of Water in Arizona." *Irrigation Age* 6 (1894): 251–253.

Responding to "Water Supplies in the Arid Regions" (**#464**), the author refutes Powell's numbers and maintains that more than double Powell's estimated 40 million acres can be irrigated.

829."The International Congress of Geologists; Reports of the American Committee to the London Session." *American Geologist* 2 (1888): 129–306.

Powell was appointed to the American Committee in 1887 but resigned in protest when the majority of the committee designated themselves final authorities on American usage of nomenclature and cartography. Powell's remarks are among those recorded in discussions of the reports.

830. James, George Wharton. "The Outspoken Hero of the Public Service: J. W. Powell." *Heroes of California: The Story of the Founders of the Golden State as Narrated by Themselves or Gleaned From Other Sources.* Boston: Little, Brown, 1910. 308–316.

James recounts the events leading up to Powell's controversial speech before the Los Angeles Irrigation Congress in 1893 (see **#439**). According to James, land speculators offered Powell a share in a "syndicate" to purchase land at a low price and buy it back from him at a high price, once the federal government passed legislation to irrigate those lands. James claims that Powell told him the story himself: "I was tempted for a while to let things go … [b]ut I could not stifle either my mind or my conscience. My duty was clear." Further, James was present at the convention for the address and describes the ensuing uproar.

831.*Limiting the Printing and Engraving for the Geological Survey.* 49th Cong., 1st Sess., H. Rept. 2740 Serial 2443. Washington: GPO, 1886.

The majority report of the Joint Commission to Consider the Present Organization of the Signal Service…, also known as the Allison Commission, supported the USGS and Powell's administration of it. See annotation for **#843**.

832. Marcou, Jules. *The Geological Map of the United States and the United States Geological Survey*. Cambridge, MA: The Author, 1892.

Geologist Marcou, a friend of Agassiz's, attacks the USGS as a politically corrupt organization and Powell as an incompetent dictator.

833.——. *A Little More Light on the United States Geological Survey*. Cambridge, MA: The Author, 1892.

Marcou again attacks Powell for mismanagement of the USGS and uncontrolled spending.

834.*Official Report of the International Irrigation Congress, Held at Los Angeles, California*. Los Angeles: Los Angeles Chamber of Commerce, 1893.

These official proceedings contain the indignant protests and comments of the delegates in response to Powell's speech, in which he argued that there was insufficient water in the West to irrigate all the land (see **#439** for Powell's speech). Also notable are remarks by Robert Brewster Stanton, in which he claims that it is feasible to build a railroad through the Grand Canyon.

835.*Report of the Special Committee of the U.S. Senate on the Irrigation and Reclamation of Arid Lands*. 51st Cong., 1st Sess., S. Rept. 928 Serial 2707 and 2708. Washington: GPO, 1890.

The final report of the committee (printed in 6 vols.) consists of two parts. The majority report, by Senator William Stewart and the Republicans, calls for irrigation to be administered by the Dept. of Agriculture. The minority report of the Democrats supports Powell's plan to keep non-irrigable lands in federal ownership for catchment basins and public pasturage. The full report was also issued separately as a "Bureau edition." See also Powell's testimony in **#414**.

836.*Report on the Sundry Civil Expenses Bill*. 47th. Cong., 1st Sess., H. Rept. 1520 Serial 2069. Washington: GPO, 1882.

Bill authorizing funds for the USGS to proceed with geologic mapping, which Powell used as an opportunity to proceed with topographic surveys.

837. Sargent, Charles Sprague. "Forests and Irrigation." *Garden and Forest* 3 (1890): 293.

Sargent was a leader of the American Forestry Association and an advocate of federal forest reservations. A critic of Powell, Sargent claimed Powell advocated the destruction of western mountain forests.

This article is a response to *The Lesson of Conemaugh* (**#390**), in opposition to Powell's view that manmade reservoirs were better than forests for water conservation.

838.——. "The Forests on the Public Domain." *Garden and Forest* 3 (1890): 13–14.

On 5 Dec. 1889, Powell delivered an address, "Problems of Irrigation," to the Chamber of Commerce in Washington (DC). Judging by the extensive quotations in this article, the text of the speech is the same as the series of articles on irrigation published in *Century Magazine* in 1890 (**#400**, **#401**, **#406**). Sargent objects to Powell's proposed "Hands off!" policy regarding governmental control of public land management in the West. In particular, Sargent argues that forests on public lands belong to the nation as much as to any local community and so should be protected as national property.

839."Sixth Annual Report of the United States Geological Survey (1884–'85)." *Science* 10 (1887): 235.

Review of the report, which "fully sustains the splendid reputation of its predecessors."

840.[Smythe, William E.]. "International Irrigation Congress." *Irrigation Age* 5 (1893): 138–145.

Smythe, editor of *Irrigation Age*, devoted an entire issue to the proceedings of the International Irrigation Congress in Los Angeles. On these pages, Smythe recounts the events of 13 Oct., when Powell shocked the delegates with his "*startling statement that there is not enough water in the arid region to irrigate one-third of the land now held under private ownership*" [italicized in original]. Contrary to telegraphed reports, Smythe asserts, Powell was not treated with discourtesy, but was allowed to explain his remarks. See **#439**, **#841**, and **#842**.

841. Smythe, William E. "Major Powell's Sensational Speech." *Irrigation Age* 5 (1893): 146–147.

In this editorial, Smythe lambastes Powell for his speech to the Irrigation Congress (**#439**). Smythe quotes from the official transcript, accusing Powell of seeking to "annihilate" the official view of the Congress that "homes for millions of men could be made on the arid public domain." See **#841** and **#842**.

842. Smythe, William E., and Elwood Mead. "The Progress of Western America." *Irrigation Age* 6 (1894): 4–5.

Smythe reports again on Powell's address to the Irrigation Congress (see **#439**). Given Powell's reputation, Smythe invites Powell to elaborate on his views in a series of articles to be published in *Irrigation Age* (see **#840**, **#841**). This article includes a letter from Elwood Mead, "A Defense of Major Powell," which defends Powell against the severity of his critics.

843.*Testimony Before the Joint Commission to Consider the Present Organization of the Signal Service, Geological Survey, Coast and Geodetic Survey, and the Hydrographic Office of the Navy Department, With a View to Secure Greater Efficiency and Economy of Operation.* 49th Cong., 1st Sess., S. Misc. Doc 82. Washington: GPO, 1886.

The Joint Commission, known as the Allison Commission for its chairman, was initiated in the July 1884 Sundry Civil Bill. Once again, the NAS was asked to submit a report on consolidating various federal bureaus and in December 1884 the commission began hearings. The resulting publication (two vols.) provides a comprehensive source of information on the state of government science in the 1880s. See Powell's testimony (**#356**) and his letter (**#342**). See also **#333**, **#334**, **#345**, and **#831**. Reprinted (**#583**).

844."The Topographical Map of the United States." *American Geologist* 10 (1892): 304–310.

This editorial opposes the continued construction of a topographical map under the authority of the USGS. and proposes that the entire topographical branch be transferred to the Dept. of the Interior. See **#827**.

2. Post-19th-Century Works

845. Alexander, Thomas G. "John Wesley Powell, the Irrigation Survey, and the Inauguration of the Second Phase of Irrigation Development in Utah." *UHQ* 37 (1969): 190–206.

Paper presented at the Utah State Historical Society's annual meeting in 1968. This is a detailed account of the period from 1888 to 1894, when conflicting interests converged in Utah over the issue of irrigation. Residents of Utah recognized the need to construct water reservoirs and welcomed the Irrigation Survey. Most were sympathetic to Powell's vision of local agricultural control over the development of irrigation systems. However, the Irrigation Survey and its surrounding political controversies instead paved the way for the transfer of political and economic control from the local to the national level, and reclamation was placed in the hands of the federal government.

846. Alexander, Thomas G. "The Powell Irrigation Survey and the People of the Mountain West." *Journal of the West* 7 (1968): 48–54.

Alexander disagrees with accounts that give the impression that people in the Mountain West opposed the Irrigation Survey. At first, they welcomed it, recognizing the need for irrigation, but opposed the way in which the survey progressed. Powell's plan to allow farmers' associations to control drainage basins "would have ... left little room for the development of manufacturing and extractive industries." As a Midwesterner, Powell placed too much emphasis on agriculture.

847. Amundson, Ronald, and Dan H. Yaalon. "E. W. Hilgard and John Wesley Powell: Efforts for a Joint Agricultural and Geological Survey." *Soil Science Society of America Journal* 59 (1995): 4–13.

Hilgard was a pioneer in soil survey. In the 1880s, Hilgard and Powell failed in their efforts to promote the transfer of the USGS from the Dept. of Interior to the Dept. of Agriculture in an attempt to implement a national soil survey.

848. Cloud, Preston. "The Improbable Bureaucracy: The United States Geological Survey, 1879–1979." *Proceedings of the American Philosophical Society* 124 (1980): 155–167.

History of the USGS and its directors.

849. Dana, Edward Salisbury, et al. *A Century of Science in America, With Special Reference to the* American Journal of Science, *1818–1918.* New Haven, CT: Yale U Pr, 1918.

Commemorates the 100th anniversary of the founding of the *American Journal of Science*. In addition to the history by Dana, subsequent chapters by prominent scientists give an overview of the history of various branches of science. George Otis Smith writes on government geological surveys, and Herbert Gregory addresses the interpretation of landforms; both discuss Powell's contributions. Good biographical information on various associates.

850. Dupree, A. Hunter. *Science in the Federal Government; a History of Policies and Activities to 1940*. Cambridge, MA: Belknap Pr and Harvard U Pr, 1957. [Rpt. NY: Harper & Row, 1964.]

> Dupree addresses Powell's leadership of the USGS, assesses the consequences of the Allison Commission's investigation in 1878 into the role of science in the federal government, and analyzes the collapse of the Irrigation Survey that led to Powell's political downfall (pp. 195–236).

851. Goetzmann, William H. "The Heroic Age of Western Geological Exploration; the U.S. Geological Survey and the Men and Events That Created It." *American West* 16.5 (1979): 4–13, 59–61.

> History of the early years of the USGS, with a focus on its publications and its foundation on the four predecessor surveys.

852. Hunt, Charles B. "John Wesley Powell, His Influence on Geology." *Geotimes* 14 (1969): 16–18.

> Hunt asserts that Powell's most important contributions to geology include his analysis of the dynamic relationship between rivers and the structural features they cross, his nomenclature of valleys, the concept of base level, and his encouragement of geological art.

853. Institute for Government Research. *The U.S. Geological Survey: Its History, Activities and Organization*. Service Monographs of the United States Government 1. New York: Appleton, 1918.

> The section on the history of the USGS, which begins with the Lewis and Clark Expedition and includes the four major western surveys, is particularly useful for its references to statutes and congressional reports that document important legal and political events. The influence of the survey's work on public land laws is discussed (pp. 59–71). Appendix 4 contains an index and compilation of authorizing laws. Appendix 6 is a bibliography of publications, articles, and manuscripts directly related to the history, activities, and organization of the survey. An important source for the early history of the USGS.

854. Manning, Thomas G. *Government in Science: The U.S. Geological Survey 1867–1894*. Lexington: U Kentucky Pr, 1967.

> A thorough, documented history of the early years of the USGS and the four predecessor surveys. An important work, with a useful bibliography and notes on archival sources.

855.——. "The Influence of Clarence King and John Wesley Powell on the Early History of the United States Geological Survey." *Interim Proceedings of the Geological Society of America, Pt. 2* (1947): 23–29.

King and Powell both supported the principle that a geological survey must support the economic interests of the nation. King saw mining as the most valuable economic force, while Powell focused on agriculture. Powell's concept of a survey was more revolutionary and clashed with tradition. An important article.

856. Merrill, George P. *The First One Hundred Years of American Geology.* New Haven, CT: Yale U Pr, 1924. [Rpt. New York: Hafner, 1969.]

A detailed chronological survey of American geology. Merrill discusses Powell's contributions to geology and his leadership of the USGS.

857. Rabbitt, Mary C. "John Wesley Powell: Pioneer Statesman of Federal Science." *The Colorado River Region and John Wesley Powell.* Geological Survey Professional Paper 669. Washington: GPO, 1969. 1–21.

This biography is primarily concerned with Powell's work as director of the USGS. Rabbitt focuses on his efforts to consolidate the geological surveys, establish the Irrigation Survey, and defend the USGS against congressional attack.

858.——. *Minerals, Lands, and Geology for the Common Defence and General Welfare, Vol. 1: Before 1879.* Washington: GPO, 1979.

In her preface to this first of a projected four-volume set, Rabbitt defines her work as a "full-scale history [of the USGS] based on primary sources ... and particularly in this volume of the geological sciences and engineering before its establishment, in terms of public-land policy, mapping policy, science policy, and the economic development of the Nation." In this context, Powell is but one of many who contribute to the formation and early development of the USGS. Rabbitt sheds light on Powell's role in Clarence King's appointment as the first director of the survey and corrects errors in previous scholarship. Includes an extensive bibliography and notes, with an index to the bibliography. See **#859**.

859.——. *Minerals, Lands, and Geology for the Common Defence and General Welfare, Vol. 2: 1879–1904.* Washington: GPO, 1980.

From the title page: "A history of geology in relation to the development of public-land, federal-science, and mapping policies and the development of mineral resources in the United States during the first 25 years of the U.S. Geological Survey." In painstaking detail, Rabbitt relays both the political and operational history of the USGS and the work of its scientists under Powell's direction (pp. 17–215). This is an essential resource for understanding the intricacies of Powell's complicated and often confrontational relationship with Congress and other federal agencies. In Rabbitt's assessment, the success of the USGS "should be attributed not to the broad view of science taken by John Wesley Powell, but to the foresight of Clarence King ... and seemingly limitless energy of Charles D. Walcott...." Extensive bibliography and notes provide a useful guide to the important government documents of the Irrigation Survey and Allison Committee. See **#858**.

860.——. *The United States Geological Survey: 1879–1989*. U.S. Geological Survey Circular 1050. Reston, VA: USGS, 1989.

Pamphlet history of the USGS.

861. Rabbitt, John C., and Mary C. Rabbitt. "The U.S. Geological Survey: 75 Years of Service to the Nation, 1879–1954." *Science* ns 119 (1954): 741–758.

The authors credit Powell with producing an unequaled body of reports and establishing the survey as an outstanding scientific body.

862. Rusho, W. L. "Powell's Hard Look at Water Facts." *Reclamation Era* 55.1 (1969): 11, 14–17.

Rusho discusses Powell's legacy regarding irrigation and reclamation in the West and quotes extensively from Powell's publications on public land policy and irrigation. Rusho acknowledges that Powell advocated only privately financed reclamation works.

863. Sageser, A. Bower. "Los Angeles Hosts an International Irrigation Congress." *Journal of the West* 4 (1965): 411–424.

Background information and highlights of the Irrigation Congress.

864. Smith, Henry Nash. "Clarence King, John Wesley Powell, and the Establishment of the United States Geological Survey." *Mississippi Valley Historical Review* 34 (1947): 37–58.

This important article brings together a wealth of primary sources to re-create the events of 1877–1878 that led to the consolidation of the four Western surveys and the establishment of the USGS. Citing congressional testimony, government publications, and unpublished letters, Smith provides a detailed account of complex political machinations.

865.——. "Rain Follows the Plow: The Notion of Increased Rainfall for the Great Plains, 1844–1880." *Huntington Library Quarterly* 10 (1947): 169–193.

In the 1860s and 1870s, Westerners created the myth that rainfall would increase as the land was cultivated for agriculture. Smith examines the relationship between this meteorological theory and the surrounding political issues and social movement of the period following the Civil War. Powell and others disproved this misconception, but it remains important as a symbol of the optimistic settlement of western lands in the 19th century.

866. Smythe, William E. *The Conquest of Arid America.* 1905. Rev. ed. Seattle: U of Washington, 1969.

Introduction by Lawrence Lee. Smythe conveys his nearly religious enthusiasm for development and settlement in the West. He extols Powell as a champion of irrigation and contributor to the reclamation of arid America. First edition 1899 by Macmillan.

867. Sperry, James E., and John Wesley Powell. "John Wesley Powell's Address at the North Dakota Constitutional Convention." *North Dakota History* 36 (1969): 369–376.

Sperry provides context for Powell's address on water rights at the North Dakota Constitutional Convention in 1889. The address is reprinted (see **#391** and **#567**).

868. Sterling, Everett W. "The Powell Irrigation Survey, 1888–1893." *Mississippi Valley Historical Review* 27 (1940): 421–434.

In this well-documented history of the establishment and dismantlement of the Powell Irrigation Survey, Sterling examines in detail the political forces at work in the House and Senate, the General Land Office, and the USGS. An essential work.

869. True, Frederick W. *A History of the First Half-Century of the National Academy of Sciences, 1863–1913.* Washington: National Academy of Sciences, 1913.

Published on the occasion of the Academy's semi-centennial. The work of the Committee on a Plan for Surveying and Mapping the Territories of the United States is summarized, with references to related articles and government publications (pp. 268–279).

870. Wallace, David Rains. *The Bonehunters' Revenge; Dinosaurs, Greed, and the Greatest Scientific Feud of the Gilded Age*. Boston: Houghton Mifflin, 1999.

The story of the "fossil feud" between paleontologists O. C. Marsh and Edward Cope and of Powell's involvement.

871. Wild, Peter. "John Wesley Powell & William E. Smythe: God Smiles on the Irrigationists." *The Opal Desert; Explorations of Fantasy and Reality in the American Southwest*. Ed. Peter Wild. Austin: U of Texas Pr, 1999. 102–109.

In this collection of essays, Wild's own piece explores how Smythe built on Powell's legacy. Powell's proposals for land reform in the West were ahead of their time, but Smythe's timing was right for his evangelical promotion of irrigation as a means to create an ideal democratic society in the West.

E. LATE WORKS: *THE CANYONS OF THE COLORADO* AND *TRUTH AND ERROR*, 1895–1902

872. Brooks, W. K. "*Truth and Error*." *Science* ns 9 (1899): 121–126.

A negative review. Brooks accuses Powell of "reifying" abstract relations and thus argues that he is guilty of the same erroneous thinking as metaphysicians, whom Powell held in contempt.

873."*Canyons of the Colorado*." *Nation* 62 (1896): 461–462.

The reviewer states that Powell's 1895 revision of *Exploration* appeals to three kinds of readers: adventure lovers and those with interests in Indians or geology

874. Fullerton, George Stuart. "Chuar, Hegel and Spencer." *Science* ns 3 (1896). 406–409.

An example of the critical response to Powell's article "Certitudes and Illusions: Chuar's Illusion" (**#489**). Fullerton objects to many of Powell's arguments and suggests that "it is not prudent for an anthropologist to venture into philosophical discussions unless he has acquainted himself with the writings of those who have preceded him."

875. McGee, WJ. "*Canyons of the Colorado*." *Science* ns 2 (1895): 593–597.

A highly favorable review of both *Canyons of the Colorado* and its author. Powell's description of the Colorado sets forth the principles of geology in a way that can be followed by the layman, the review contends.

876.——. "The Foundations of Science." *Forum* 27 (1899): 168–178.

Part I is a review of Lord Bacon's *Novum Organum*. Part II is a review of *Truth and Error*, which McGee sees as an extension of Bacon's work. He concludes that the "old organon and the new appear as the modern foundation of science," but admits that Powell's work may lie dormant.

877."*Truth and Error, or the Science of Intellection*." *The Critic* 34 (1899): 375–376.

A favorable review. "This curious work" will be of interest to the "expert reader."

878. Ward, Lester Frank. "Review of *Truth and Error, or the Science of Intellection....*" *Glimpses of the Cosmos*. Vol. 6. New York: G.P. Putnam's Sons, 1913. 51–54.

Powell's good friend defends his decision to criticize *Truth and Error*, even though Powell dedicated it to him. Powell never let the review come between their friendship. See **#153**, **#510**, and **#879**.

879.——. "*Truth and Error*." *Science* ns 9 (1899): 126–137.

Despite Powell's dedication of *Truth and Error* to Ward, Ward's review is not favorable. He criticizes Powell for his excessive introduction of new terms and fixation on "pentologies" and argues that Powell tries too hard to find a logical order in chaotic Nature. See **#510** and **#878**.

Chapter 4

Works About John Wesley Powell's Associates

880. Anderson, Martin J. "Artist in the Wilderness: Frederick Dellenbaugh's Grand Canyon Adventure." *Journal of Arizona History* 28 (1987): 47–68.

A biography of Dellenbaugh, with a focus on the artist's experience during the second descent of the Colorado River by the Powell Survey. Anderson relays how Dellenbaugh's feelings changed through the years as his admiration for Powell turned to total adoration. Dellenbaugh exaggerated or modified facts and committed hundreds of errors when he dictated A. H. Thompson's diary to a stenographer in 1929.

881. Barrows, Eliza J. Howland. "Howland Brothers Join the John Wesley Powell Expedition to Explore the Colorado River That Flows Through the Grand Canyon—April 1869." *Howland Quarterly* 50.4 (1986): 2–4.

Published by the Pilgrim John Howland Society. Barrows tells the story of the 1869 expedition through Separation Rapid from the fictionalized first-person voice of Powell. The deaths of the Howland brothers and Dunn are blamed on Shivwits Indians.

882. Chamberlain, Ralph V. "Francis Marion Bishop, 1843–1933." *UHQ* 15 (1947): 155–158.

A biographical sketch of the mapmaker for the second river expedition.

883. Darrah, William Culp. "Beaman, Fennemore, Hillers, Dellenbaugh, Johnson and Hattan." *UHQ* 16–17 (1948–1949): 491–503.

Biographical sketches of six men associated with the second Colorado River expedition: E. O. Beaman, James Fennemore, John K. Hillers, Frederick S. Dellenbaugh, William D. Johnson, and Andrew Hattan.

884.——. "George Young Bradley, 1836–1855." *UHQ* 15 (1947): 29–30.

Biographical sketch of the most reliable diarist of the 1869 Colorado River expedition.

885.——. "Hawkins, Hall and Goodman." *UHQ* 15 (1947): 106–108.

Biographical sketches of three members of the first Colorado River expedition: Frank Goodman, Andy Hall, and Billy Hawkins.

886.——. "The Howland Brothers and William Dunn." *UHQ* 15 (1947): 93–94.

Biographical sketches of Oramel, his brother Seneca, and William Dunn, who left the 1869 river party at Separation Rapid.

887.——. "John C. Sumner, 1840–1907." *UHQ* 15 (1947): 109–112.

Brief biography of Jack Sumner and the controversy surrounding his journals (see **#56**, **#57**, **#137**, **#138**, **#140**).

888.——. "John F. Steward." *UHQ* 16–17 (1948–1949): 175–179.

A biographical sketch with information about Steward's field notes from the 1871–1872 expedition (see **#54**).

889.——. "Walter Henry Powell, 1842–1915." *UHQ* 15 (1947): 89.

A brief sketch of Powell's youngest brother, who accompanied Powell on the 1868 and 1869 expeditions.

890. Dawson, Thomas F. "Lost Alone on Bear River Forty Years Ago." *The Trail* 11.2 (1918): 13–20.

Based on an interview, this is the story of student William H. Bishop (not to be confused with Francis M. Bishop) and his experience with Powell in Colorado in 1868.

891. Euler, Robert C. "Frederick Dellenbaugh: Grand Canyon Artist." *Journal of Arizona History* 28 (1987): 31–46.

Biography, illustrated with several of Dellenbaugh's paintings of the Grand Canyon.

892. Fernlund, Kevin J. *William Henry Holmes and the Rediscovery of the American West*. Albuquerque: U of New Mexico Pr, 2000.

This biography of survey artist Holmes includes details of the field trip to Cuba with Powell and S. P. Langley in 1900.

893. Foster, Mike. *Strange Genius: The Life of Ferdinand Vandeveer Hayden.* Niwot, CO: Roberts Rinehart, 1994.

A biographical investigation into Hayden's fall from the most prominent of the four western surveyors in 1879 to the most neglected. Powell comes in for serious criticism as a political manipulator and self-promoter who has been overly romanticized by his biographers. This biography is strong on the events surrounding the formation of the USGS and King's appointment to director.

894. Fowler, Don D. *The Western Photographs of John K. Hillers: "Myself in the Water."* Washington: Smithsonian Institution Pr, 1989.

Richly illustrated with many photographs taken by Hillers during his years with the Powell Survey and the USGS, this monograph describes his photographic opus (1871–1900). An appendix contains corrections and comments on the content of two catalogs of Hillers' photographs.

895. Gregory, Herbert E. "Stephen Vandiver Jones." *UHQ* 16–17 (1948–1949): 10–18.

Sketch of a fellow Illinoisan who kept a journal of the second river expedition.

896.——. "Walter Clement Powell." *UHQ* 16–17 (1948–1949): 253–255.

Biographical sketch of cousin "Clem," who accompanied the 1871–1872 expedition as a photographer.

897. King, Clarence, et al. *Clarence King Memoirs; The Helmet of Mambrino.* New York: G. P. Putnam's Sons, 1904.

The King Memorial Committee of the Century Club honored King by reprinting his short story "The Helmet of Mambrino" along with a number of personal memoirs by friends and associates.

898. Kinsey, Joni L. *Thomas Moran and the Surveying of the American West.* New Directions in American Art. Washington: Smithsonian Institution Pr, 1992.

Using Moran's paintings and drawings of Yellowstone and the Grand Canyon, Kinsey demonstrates how art influenced America's perception of the West. Moran's paintings of the Canyon and Powell's reports present a shared vision of the grandeur and awesome story of the landscape. The rhythmic cadences and symbolic allusions in Powell's writing "have their source not only in the Bible but also in the land itself." A thorough study with extensive notes and bibliography.

899. Lago, Don. "Billy Hawkins: Abducted by Alias!" *boatman's quarterly review* 14.3 (2001): 10–11.

Hawkins (a.k.a. Billy Rhodes) accompanied Powell in 1869. Lago corrects errors in Darrah's research on Hawkins' biography. See **#885**.

900. Lessard, F. Dennis. "E. O. Who?" *American Indian Art Magazine* 12.2 (1987): 52–61.

In this study of Beaman's photographic works, Lessard notes that Powell discredited Beaman's work when he left the expedition in 1872. However, his work contributed to the careers of both Powell and John Hillers.

901. Masteller, Richard N. "Western Views in Eastern Parlors: The Contribution of the Stereograph Photographer to the Conquest of the West." *Prospects: An Annual of American Cultural Studies* 6 (1981): 55–71.

The stereographic images of E. O. Beaman and John Hillers provided Easterners with information about the Green and Colorado Rivers and reminded Americans that western lands were theirs. Masteller argues that they simultaneously devalued the land because the images were inexpensive and readily available.

902. Olsen, Robert W., Jr. "Clem Powell and Kanab Creek." *The Kiva* 34 (1968): 41–50.

Olsen describes two occasions when Walter Clement Powell was in Kanab Creek Canyon while working as a photographer for the Powell Survey in 1872.

903. Osborn, Henry Fairfield. *Cope: Master Naturalist; the Life and Letters of Edward Drinker Cope With a Bibliography of His Writings Classified by Subject.* Princeton, NJ: Princeton U Pr, 1931. [Rpt. New York: Arno Pr, 1878.]

Biography by an ally of Cope. Of interest are letters to friends and family, which give Cope's view of his antagonistic relationship with Powell.

904. Pyne, Stephen J. *Grove Karl Gilbert.* Austin: U of Texas Pr, 1980.

This is a comprehensive biography of G. K. Gilbert, who began his 30-year collaboration with Powell in 1875 when he joined the Powell Survey. Pyne notes that the two men were like brothers, despite their different temperaments. Gilbert was a superb scientist and explorer, the antithesis of the romantic. In contrast, Powell "is a romantic naturalist in the vogue of Newberry." Powell's contributions to geology were a few dramatic insights that generated complex systems of classifications.

905. Resek, Carl. *Lewis Henry Morgan, American Scholar*. Chicago: U of Chicago Pr, 1960.

Biography of the founder of American anthropology and mentor to Powell.

906. Schuchert, Charles, and Clara Mae LeVene. *O. C. Marsh, Pioneer in Paleontology*. New Haven, CT: Yale U Pr, 1940.

Biography by a proponent of Othniel Marsh, who was Powell's ally as president of the National Academy of Science and head of the USGS Division of Paleontology.

907. Stegner, Wallace. "C. E. Dutton: Explorer, Geologist, Nature Writer." *Scientific Monthly* 45 (1937): 81–85.

Dutton served with Powell as a geologist for the Rocky Mountain Survey and the USGS.

908. Watson, Elmo Scott, and Julia Watson. "The Major's Right-Hand Woman." *Empire Magazine* (*Denver Post*) 59.298 (May 27, 1951): 30–31.

The Watsons glean newspaper articles and other secondary sources to trace the path of Powell's wife, Emma Dean, who accompanied him on both expeditions to the Rocky Mountains (1867 and 1868) and Utah in 1871–1872.

909. Wilkins, Thurman, and Caroline Lawson Hinkley. *Clarence King: A Biography*. Rev. and enl. ed. Albuquerque: U of New Mexico Pr, 1988.

Biography of the head of the Fortieth Parallel Survey and first director of the USGS. Wilkins agrees with Mary Rabbitt's assessment that the success of the USGS was due not to Powell, but to King's "broad view of science." See **#859**.

910. Wilkins, Thurman, and Caroline Lawson Hinkley. *Thomas Moran: Artist of the Mountains*. 2nd ed., rev. and enl. Norman: U of Oklahoma Pr, 1998.

Scholarly biography of the artist chosen by Powell to illustrate (and promote) his expeditions to Utah and Arizona in the 1870s.

Chapter 5

Historiography

A. HISTORICAL AND LITERARY ASSESSMENTS, INCLUDING CENTENNIAL (1869–1969) PUBLICATIONS

911. Aton, James M. *Inventing John Wesley Powell: The Major, His Admirers, and Cash-Register Dams in the Colorado River Basin.* Distinguished Faculty Lecture 9. Cedar City: Southern Utah State College, 1988.

> Using Powell and the federal program of dam construction on the Colorado River as his subject, Aton examines Powell's writings on reclamation and explores how these writings have been interpreted and reinterpreted. Powell has been reinvented as a conservationist, an environmentalist, a conquering frontiersman, and more, by biographers Stegner, Dellenbaugh, and others according to their bias. Aton's Powell abhorred big industrial power and might have become an environmentalist in our time.

912.——. *John Wesley Powell.* Boise State University Western Writers Series 114. Boise, ID: Boise State U, 1994.

> Aton presents a short biography of Powell followed by literary analyses of *Exploration, Report on the Lands,* and other publications. Joseph Campbell's *The Hero with a Thousand Faces* informs Aton's analysis of *Exploration.* He notes Powell's "Nature-as-Book" metaphor and use of figures of speech from 19th-century Romanticism. Regarding his work in anthropology, Powell was a good fieldworker, as evidenced by Fowler's *Anthropology of the Numa* (**#29**). The published works of the BAE produced landmark publications in Native American studies.

913. Bartlett, Richard Adams. "John Wesley Powell and the Great Surveys: A Problem in Historiography." *American West: An Appraisal.* Ed. Robert G. Ferris. Santa Fe: Museum of New Mexico Pr, 1963. 48–57.

> According to Bartlett, historians must sometimes destroy a "personality image" to present accurate history. This is the case with Powell, who

was lionized at the expense of other explorers, resulting in "an inaccurate picture of government-sponsored scientific exploration in the American West." Each of the other surveys made strong contributions to science and government service, but they are underrepresented in history texts.

914. Box, Thadis W. *The Arid Lands Revisited—One Hundred Years After John Wesley Powell.* Annual Faculty Honor Lecture 57. Logan: Faculty Association, Utah State U, 1978.

Box assesses the condition of the arid lands a century after the publication of *Report on the Lands* After discussing Powell's plan for the management of land in the arid West, Box presents statistical evidence to show that the potential that Powell saw for the West has not been achieved. Western lands are being controlled by the views of "the people from the humid East." Box proposes legislative and administrative reforms to promote land-use planning and assist private landowners.

915. Bryson, Michael A. "Controlling the Land: John Wesley Powell and the Scientific Management of the American West." *Science, Values, and the American West.* Ed. Stephen Tchudi. Reno: Nevada Humanities Committee, 1997. 3–23.

A critique of Powell's method and rhetoric in *Report on the Lands* Powell uses quantitative analysis for his rhetorical framework; science allows Powell to construct nature as an active entity, which can in turn be controlled by science for human use. Bryson also addresses Powell's views of the native peoples who, within the scope of his *Report*, "embody an ecological problem to be overcome." Ultimately, *Report* is a "sophisticated attempt to reformulate the relationships among nature, science, and human society." See **#916**.

916.——. "'A Unit of Country Well Defined in Nature;' John Wesley Powell and the Scientific Management of the American West." *Visions of the Land: Science, Literature, and the American Environment From the Era of Exploration to the Age of Ecology*. Charlottesville: U Press of Virginia, 2002. 80–101.

Essentially the same as "Controlling the Land" (**#915**) with some revision.

917. Carter, Edward Carlos. *Surveying the Record: North American Scientific Exploration to 1930*. Memoirs of the American Philosophical Society 231. Philadelphia: American Philosophical Society, 1999.

A collection of papers presented at a conference hosted by the APS. Historians, anthropologists, art historians, and others examined historical approaches to scientific expeditions and surveys. Topics reflect the various fields of participants, such as anthropology by Don Fowler and David Wilcox. The last essay, by Donald Worster, is on Powell (see **#791** for annotation).

918.*Colorado River Explorer Powell; Centennial Issue: 1869–1969*. Spec. issue of Reclamation Era 55.1 (1969).

Contents include: "Centennial to Honor Major Powell," by Gordon J. Forsyth (**#923**); "Famous Trip Boosts Powell's Career," by Mary C. Rabbitt (**#767**); and "Powell's Hard Look at Water Facts," by W. L. Rusho (**#862**).

919.*The Colorado River Region and John Wesley Powell*. Geological Survey Professional Paper 669. Washington: GPO, 1969.

"A collection of papers honoring Powell on the 100th anniversary of his exploration of the Colorado River." Includes a biography of Powell by Rabbitt (see **#857** for annotation). The other three essays are technical papers focusing on the geology of the Grand Canyon.

920. Davison, Stanley Roland. *The Leadership of the Reclamation Movement, 1875–1902*. Development of Public Land Law in the United States. New York: Arno Pr, 1979.

This work is critical of large, federal irrigation projects for reclamation and the men who organized the reclamation movement. They saw irrigation as the single cure for all the troubles of the West, regardless of its economic practicality. Powell has come to be regarded as the reclamation movement's first leader; however, though he had an important part in effecting interest in national policy, he parted ways with the reclamationists at the International Irrigation Congress in 1893. Davison's focus is on Powell's politics regarding irrigation, but in the process he analyzes Powell's personality and critiques some of his later writing.

921.deBuys, William. "Saint Contrary: John Wesley Powell." *High Country News* 31.7 (1999).

Powell was a bold and unconventional thinker who saw the West "on its own terms, as a place that needed to grow its institutions and wrestle with its problems in its own homegrown way." Had his commonwealth ideals taken hold, Washington's "administrative pyramids" would have toppled. Journal available online at <www.hcn.org>.

922. Dimock, Brad. "White Lies, Major Distortions, and Several New Stabs at the Truth—8 Recent Publications on James White and Major John Wesley Powell." *boatman's quarterly review* 15.1 (2002): 20–26.

A comparative review of three books about James White and five about Powell: *A River Running West,* by Donald Worster (**#689**); *Down the Great Unknown,* by Edward Dolnick (**#742**); *Seeing Things Whole,* by William deBuys (**#213**); *The Last Canyon,* by John Vernon (**#1054**); and *"The River Is Mine,"* by Ardian Gill (**#1051**).

923. Forsyth, Gordon J. "Centennial to Honor J. W. Powell." *Reclamation Era* 55.1 (1969): 1–5.

Overview of special events and exhibits planned for the centennial of Powell's first Colorado River expedition.

924. Four Corners Geological Society. *Geology and Natural History of the Fifth Field Conference, Powell Centennial River Expedition 1969.* Durango, CO: Four Corners Geological Society, 1969.

This collection of papers was used as background material for a conference organized and sponsored by the Four Corners Geological Society to commemorate the centennial of Powell's famous expeditions down the Colorado River. Most papers are devoted to the geology, geography, and natural history of the Grand Canyon and lower Colorado River. See **#595**.

925. Gabriel, Ralph Henry. *The Course of American Democratic Thought; An Intellectual History Since 1815.* 2nd ed. New York: Ronald Pr, 1956.

In his study of social beliefs of 19th-century America, Gabriel examines the influence of Darwin, Lewis Henry Morgan, and Powell (pp. 170–182).

926. Hess, Karl Jr. "John Wesley Powell and the Unmaking of the West." *Environmental History* 2.1 (1997): 7–28.

Slightly revised version of the essay in *The Next West.* See **#927** for annotation.

927.——. "John Wesley Powell and the Unmaking of the West." *The Next West: Public Lands, Community, and Economy in the American West.* Ed. John A. Baden and Donald Snow. Washington: Island Pr, 1997. 151–180.

According to Hess, The American West of today is environmentally degraded because of men like Powell, who tried to re-create eastern agrarianism through the imposition of federal government control.

Wallace Stegner's Powell, as imagined in *Beyond the Hundredth Meridian* (**#671**), was not a social and environmental visionary. Instead, he is responsible for damming rivers, cutting primeval forests, and ruining rangelands. See **#955** for Worster's response.

928. Hutchinson, Charles F. "John Wesley Powell and the New West." *COSMOS Journal* 10 (2000). <www.cosmos-club.org/journals/2000/hutchinson.html>.

Powell's revolutionary recommendations regarding land management for the West are being embraced in the "New West," according to Hutchinson. Community models of decision making are appearing to determine not only water rights, but also recreational and other types of land use.

929.*John Wesley Powell and the Colorado River, Centennial Edition.* Spec. issue of *UHQ* 37 (1969): 146–283.

Commemorating the 1869 expedition. Articles by William Darrah, W. L. Rusho, and others are cited individually (see **#110**, **#140**, **#607**, **#762**, **#770**, **#803**, **#845**).

930. Judge, Joseph, and Walter M. Edwards. "Retracing John Wesley Powell's Historic Voyage Down the Grand Canyon." *National Geographic* 138 (1969): 688–713.

Judge and his party commemorate the centennial of the 1869 expedition by rafting the Colorado River through Marble and Grand Canyons. Illustrated with photographs by Edwards.

931. Kelly, Charles. "The Three Who Lost." *Desert Magazine* 8.6 (1945): 4–6.

In 1939, Julius Stone placed a plaque in the Grand Canyon that marks the events at Separation Rapid. Kelly retells the incident and gives Stone's rationale for supporting Stanton's version in *Colorado River Controversies* (**#774**).

932. Kemmis, Daniel. *This Sovereign Land: A New Vision for Governing the West*. Washington: Island Pr, 2001.

Examines today's applications of Powell's ideas regarding public land policy.

933. Kirsch, Scott. "John Wesley Powell and the Mapping of the Colorado Plateau, 1869–1879: Survey Science, Geographical Solutions, and the Economy of Environmental Values." *Annals of the Association of American Geographers* 92 (2002): 548–572.

> The mapping of geographical spaces and boundaries, as exemplified in the cartographic and land classification work of the Powell Survey and the census and survey work undertaken by Powell as a special commissioner for the Bureau of Indian Affairs, provided "spatial knowledge" by which the government could impose its values of efficiency, order, and productivity on western lands.

934.——. "Regions of Government Science: John Wesley Powell in Washington and the American West." *Endeavour* 23 (1999): 155–158.

> Kirsch examines the formation and impact of Powell's career in terms of his role as a scientific explorer of the arid regions of the American West and his involvement with the emerging community of government science in Washington.

935. Maxwell, John. "John Wesley Powell and Democracy: A Textual Politics for the Arid Region." *Journal of the Southwest* 37 (1995): 482–494.

> Contrary to widely held opinions favorable to Powell, Maxwell believes Powell contributed to the unequal distribution of water and land to Indians. While not as exploitative as his contemporaries, Powell's values and attitudes, evidenced in his writing, discriminate against Indians. He wanted to remove Indians from the influence of the lowest class of white men to reservations where they could "safely make the transition from their own practices to those of civilization."

936. Nelson, Clifford M. "*A River Running West: The Life of John Wesley Powell.*" *Geotimes* 46.11 (2001): 52–53.

> In this review of Donald Worster's biography of Powell (see **#689**), Nelson disagrees with Worster's assessment of Powell as an administrator. Unlike Clarence King, Powell did a poor job as director of the USGS because he was not responsive to the public interest.

937.O'Callaghan, Jerry A. "John Wesley Powell and the Public Lands." *Our Public Lands* 19.3 (1969): 6–9.

> O'Callaghan contends that Powell's proposals for the systematic classification of public lands led to later federal acts and policies for the administration and classification of coal, oil and mineral lands.

938. Opie, John. "John Wesley Powell Was Right: Resizing the Ogallala High Plains." *Fluid Arguments: Five Centuries of Western Water Conflict*. Ed. Char Miller. Tucson: U of Arizona Pr, 2001. 223–247.

Opie proposes changing the basis of map information from counties and sections to natural resources, as Powell first proposed in *Report on the Lands*.

939. Pisani, Donald J. "Forests and Reclamation, 1891–1911." *Forest & Conservation History* 37 (1993): 68–79.

This article looks at the ebb and flow of conflict and resolution between the goals and policies of the Forest Service and the Reclamation Service. Powell's *Report on the Lands* threatened forest preservation by creating irrigation and grazing districts at the expense of the woodlands. Powell was concerned that forest preserves would lock up reservoir sites and that new bureaucracies would compete with the USGS for funds.

940.——. *To Reclaim a Divided West: Water, Law, and Public Policy, 1848–1902*. Histories of the American Frontier. Albuquerque: U of New Mexico Pr, 1992.

In this scholarly approach to the complex issues surrounding western water laws, Pisani analyzes the organizations that advocate national and state programs and studies the lives of those who promoted specific programs dealing with water resources. Regarding Powell (pp. 127–168), Pisani's position is that he never favored a national program of reclamation. Rather, Congress lost faith in the Irrigation Survey because Powell refused to participate in Western boosterism by advocating federal irrigation projects. Pisani concludes that Powell's biographers mistakenly portray him as a rational idealist who battles corrupt capitalists (Worster), or as a vain, self-serving egomaniac (Davison). Good bibliography and notes.

941. Pyne, Stephen J. *Dutton's Point: An Intellectual History of the Grand Canyon*. Monograph 5. Grand Canyon, AZ: Grand Canyon Natural History Association, 1982.

Pyne discusses the explorers, scientists, artists, writers, and intellectuals have who experienced, interpreted, and shaped the public perception of the Grand Canyon. Regarding Powell's portrayal of the Canyon in *Exploration*, he writes in the tradition of the explorer as romantic hero, not as a scientist. Powell's experience on the Colorado River "excited Powell's imagination ... [and] would animate his subsequent career as a geologist and administrator of the Geological Survey." The author's introduction notes that "portions of this book are taken from *Grove*

Karl Gilbert" (see **#904**). Revised and reprinted with the new title *How the Canyon Became Grand* (see **#942**).

942.——. *How the Canyon Became Grand; A Short History*. New York: Penguin, 1998.

A revised, expanded version of *Dutton's Point* (**#941**). An appendix contains graphs and charts illustrating the history of exploration in the Grand Canyon, annual visitation, the output of Canyon-related publications, and more.

943. Sarles, Frank B. *John Wesley Powell and the Colorado River; a Special Study of the Colorado River Expeditions of 1869 and 1871*. N.p.: National Survey of Historic Sites and Buildings Office of Archeology and Historic Preservation, U.S. Dept. of the Interior, National Park Service, 1968.

Following an introductory biography and history of both expeditions, Sarles relays his own journey following Powell's route down the Green and Colorado Rivers. Appended are a list of national parks and historic sites along the route and unnumbered pages of maps and photographs.

944. Smith, Henry Nash. *Virgin Land: The American West as Symbol and Myth*. New York: Random House/Knopf, 1950. [Rpt. Cambridge, MA: Harvard U Pr, 1970.]

Smith traces the impact of western settlement on American thought and literature. In "The Myth of the Garden and Reform of the Land System" (pp. 195–200), he examines the 1878–1879 congressional battle over Powell's proposal for settlement of the arid lands. Powell's reforms were too drastic: "The dream of a yeoman society" did not yield to rationalism and science. A standard work on the intellectual history of the West and one that reintroduced Powell to historians.

945. Stegner, Wallace Earle. "Getting to Know the National Domain." *American Heritage* 30.2 (1979): 60–62.

Powell is praised for his direction of the USGS and the BAE, which a hundred years later still provide reliable information to the public and remain as useful, respected bureaus of the federal government.

946. Stephens, Hal G., and Eugene Merle Shoemaker. *In the Footsteps of John Wesley Powell: An Album of Comparative Photographs of the Green and Colorado Rivers, 1871–'72 and 1968*. Boulder, CO: Johnson Bks, 1987.

A selection of 110 of the 150 "camera stations" of the Powell expeditions, with photographs by Hillers and others juxtaposed with full-color photographs by Stephens and accompanied by essays on the sites by Shoemaker. The focus is the geography and geology of the Colorado River canyons.

947. Van Noy, Rick. *Surveying the Interior: Literary Cartographers and the Sense of Place.* Environmental Arts and Humanities Series. Reno: U of Nevada Pr, 2003.

A study of the writings of four surveyors and mappers as literary cartographers: Thoreau, Clarence King, Powell, and Wallace Stegner. In "Surveying the Sublime: John Wesley Powell's Representations of Place" (pp. 100–141), Van Noy demonstrates that while Powell proposed "a more 'realistic mapping' of the West," he described the landscape in romantic (heroic and sublime) terms. His analysis focuses on *Exploration, Canyons of the Colorado*, and *Report on the Lands.*

948. Webb, Walter Prescott. *The Great Plains.* Boston: Ginn, 1931. [Rpt. Lincoln: U of Nebraska Pr, 1981.]

This is one of the earliest histories of the region to recognize the importance of aridity in the development of the American West, and Webb is the first modern historian to rediscover Powell. While he sees Powell's ideas, as expressed in *Report on the Lands* and "Institutions for the Arid Lands" (**#400**), as the foundation for large-scale federal reclamation projects of the 20th century, Webb asserts that "Powell nowhere recommended that the national government should participate in reclamation."

949. White, Richard. "Nature or Justice." *The New Republic* 224.24 (2001): 47–52.

This essay contrasts the portrayal of Powell in the biographies by Donald Worster (**#689**) and Wallace Stegner (**#671**). White discusses Powell's inability to "forge nature and justice into an inseparable whole" and analyzes Worster's approach to the issue in *A River Running West.*

950. Worster, Donald. *Encountering Mormon Country: John Wesley Powell, John Muir, and the Nature of Utah.* Leonard J. Arrington Mormon History Lecture Series 8. Logan: Special Collections and Archives, Utah State U, 2003.

Comparing Powell and Muir, Worster discusses their impressions of Utah and the impact of Mormon society on their thinking about the relationship of man to his environment in the West.

951.——. "Landscape With Hero: John Wesley Powell and the Colorado Plateau. " *Southern California Quarterly* 79.1 (1977): 29–46.

Worster views Powell as a complex man who has been claimed and reinvented in contradictory ways. To understand Powell is to understand the diverse ways that Westerners have reacted to their natural environment. Worster maintains that the voyage through the canyons of the Colorado had a "liberating effect" on Powell and those that followed.

952.——. "The Legacy of John Wesley Powell." *An Unsettled Country: Changing Landscapes of the American West.* Calvin P. Horn Lectures in Western History and Culture. Albuquerque: U of New Mexico Pr, 1994. 1–30.

Worster sees Powell not as a champion of federal land management, but as a proponent of local land ownership and governance. Powell's chief contribution was his recognition of the importance of aridity in the settlement and development of American West. Revised and republished in *Old West New West* (**#954**) and *Reopening the American West* (**#953**).

953.——. "The Legacy of John Wesley Powell." *Reopening the American West.* Ed. Hal K. Rothman. Tucson: U of Arizona Pr, 1998. 77–89.

Slight revision of **#952**. See **#954**.

954.——. "Rediscovering the West, the Legacy of John Wesley Powell." *Old West/New West: Quo Vadis?* Ed. Gene M. Gressley. Worland, WY: High Plains Pub, 1994. 101–122.

Expanded, revised version of "The Legacy of John Wesley Powell" (see **#952**, **#953**). Ending on a more pessimistic note, Worster calls for a history of the American West that "builds on Powell's early insights into the significance of the land." Since Powell could not have foreseen today's environmental destruction, a full environmental history must be written and learned from in order to move "far beyond Powell."

955.——. "A Response to "John Wesley Powell and the Unmaking of the American West"." *Environmental History* 2 (1997): 216–219.

Worster rebukes Karl Hess for his sloppy scholarship, conservative ideology, and distortion of Powell's views. See **#927** for annotation of Hess' article.

956.——. *Rivers of Empire: Water, Aridity, and the Growth of the American West.* New York: Pantheon Bks, 1985.

Worster contends that water is the driving force in past and future settlement and development of the American West. Powell opposed the control of water by capitalists or national government, opting instead for self-determination within water districts. His plan for settlement had workable features, but there were serious internal and external problems that Powell never addressed.

957. Yochelson, Ellis Leon. "John Wesley Powell, the Monuments." *Geotimes* 14 (1969): 19–20.

A survey of historical signs, markers, monuments, and place-names that honor Powell. Similar to **#958**.

958.——. "Monuments and Markers to the Territorial Surveys." *Annals of Wyoming* 43.1 (1971): 113–124.

A survey of "tangible memorials" to King, Powell, Hayden, and Wheeler. Expansion of **#957**.

959. Young, John V. "In the Wake of John Wesley Powell." *Westways* 61.6 (1969): 26–29, 50.

Compares modern river rafting on the Colorado River to the experience of Powell's party a century before.

960. Zwinger, Ann. "'... A Worthless and Impracticable Region....'" *Plateau* 52.2 (1980): 24–32.

An examination of the historical significance of the confluence of the Green and Colorado Rivers and the explorers who wrote about it—Newberry, Powell, and Stanton.

B. BIBLIOGRAPHIES AND GUIDES

961. Burger, Barbara Lewis, comp. *Guide to the Holdings of the Still Picture Branch of the National Archives.* Washington: NARA, 1990.

Excellent guide to photographs in various NARA collections.

962.*Catalogue of Numbered Manuscripts at the National Anthropological Archives.* Boston: G. K. Hall, 1975.

Photocopy of card index. Records are updated and available in the Smithsonian online library catalog at <www.si.siris.edu>.

963. *The Colorado River, People & Places; A Catalogue of Books, Pamphlets, Maps and Manuscripts*. Scottsdale, AZ: Guidon Bks, 1969.

This is a sale catalog for Guidon Books, "issued on the centennial of J. W. Powell's voyage discovering the Colorado River." Introduction by C. Gregory Crampton. Contains books, ephemera (such as postcards and stereopticon views), and journals.

964. Farquhar, Francis P. *The Books of the Colorado River & the Grand Canyon, a Selective Bibliography*. Los Angeles: Glenn Dawson, 1953. [Rpt. Flagstaff, AZ: Fretwater Pr, 2003.]

Beautifully printed. A classic annotated bibliography of important publications on the region. See **#965**.

965. Ford, Mike S. *The Books of the Grand Canyon, the Colorado River, the Green River & the Colorado Plateau 1953–2003; a Selective Bibliography*. Flagstaff, AZ: Fretwater Pr, 2003.

Update of Farquhar's annotated bibliography (**#964**).

966. Glenn, James R. *Guide to the National Anthropological Archives Smithsonian Institution*. Rev. and enl. ed. Washington: NAA, 1996.

An essential resource.

967. Goodman, David Michael. *Arizona Odyssey: Bibliographic Adventures in Nineteenth-Century Magazines*. Tempe: Arizona Historical Foundation, 1969.

Bibliography of magazine articles about Arizona published 1850–1900.

968. *Guide to Cartographic Records in the National Archives*. Washington: NARA General Services Administration, 1971.

Although outdated, useful for identifying map collections in the NARA.

969. *Guide to the Smithsonian Archives 1996*. Archives and Special Collections of the Smithsonian Institution 5. Washington: Smithsonian Institution Pr, 1996.

This thick volume provides detailed descriptions of the collections and helpful indexes and appendices.

970. Hodge, Frederick Webb. *List of the Publications of the Bureau of Ethnology*. BAE Bulletin 24. Washington: GPO, 1894.

Contains all the major works of the BAE's first decade.

971. Kvasnicka, Robert M. *The Trans-Mississippi West, 1804–1912; Part IV, A Guide to the Records of the Department of the Interior for the Territorial Period; Section 2, Records of Select Agencies*. Washington: NARA, 1997.

Contains an excellent section on USGS record groups.

972. *List of Publications of the Bureau of American Ethnology With Index to Authors and Titles*. BAE Bulletin 49. Washington: GPO, 1910.

An essential bibliography. Second impression published in 1911.

973. Lurie, Edward. "Some Manuscript Resources in the History of Nineteenth Century American Natural Science." *Isis* 44 (1953): 363–370.

A survey of manuscript collections.

974. Matchette, Robert B. *Guide to Federal Records in the National Archives of the United States*. Washington: NARA, 1995.

Regularly updated Web version of the *Guide* available online at <www.nara.gov/guide>.

975. Nickles, John M. *Geologic Literature on North America 1785–1918*. USGS Bulletin 746. Washington: GPO, 1923.

Especially useful for its indexing of International Congress of Geologists publications.

976. Schmeckebier, Lawrence Frederick. *Catalogue and Index of the Publications of the Hayden, King, Powell, and Wheeler Surveys*. USGS Bulletin 222: Series G, Miscellaneous 26. Washington: GPO, 1904.

An essential source for publications of the predecessor surveys. The works listed for the Powell Survey do not include any of Powell's reports published before 1876, when the U.S. Geographical and Geological Survey of the Rocky Mountain Region was named and appropriated. Also issued as 58th Cong., 2nd Sess., H. Misc. Doc. 606.

977. Spamer, Earle E. *Bibliography of the Grand Canyon and Lower Colorado River*. Internet (3rd) ed. Grand Canyon Association, 2003. <www.grandcanyon.org/biblio>.

A remarkable, comprehensive online bibliography for anything and everything published about or related to the Colorado River region since 1540. Subjects range from geological surveys to poetry, and formats range from print to audio to computer media. For anyone with an interest in Powell at any level; the researcher will find additional sources not included in the scope of this annotated bibliography (such as history books with only passing mention of Powell or brief excerpts in anthologies). The website has a sophisticated search engine, allowing for easy search and retrieval of citations. Continually updated.

978. Walsworth, Bertha L., comp. *The Colorado River and Its Tributaries; a Bibliography of Books Magazines Articles and Government Documents in the Riverside Public Library*. Riverside, CA: Riverside Public Library, 1922.

Typed manuscript of an annotated bibliography. Many citations relate to water and reclamation projects, but the manuscript also includes earlier explorations and travel articles. A supplement was added in 1926 by Florence S. Hellman at the Library of Congress, titled *List of References on the Colorado River and Its Tributaries.*

979. Warman, Philip C. *Catalogue and Index of the Publications of the United States Geological Survey, 1880–1901*. USGS Bulletin 177. Washington: GPO, 1901.

A comprehensive bibliography of all works published by the USGS. See also **#976**.

980.——. "Catalogue of the Published Writings of John Wesley Powell." *Proceedings of the Washington Academy of Sciences* 5 (1903): 131–187.

For a century, Warman's article stood as the only nearly complete bibliography of Powell's published works, though it contains errors, omissions, and duplications. Warman's work was apparently based on publications held in the Smithsonian's Dept. of Anthropology Library, at some point bound in ten volumes and titled *Published Writings of John Wesley Powell* (**#16**). In "John Wesley Powell: Proceedings of a meeting ..." (**#640**).

Chapter 6

Dissertations and Theses

981. Bartlett, Richard Adams. "The Great Surveys in Colorado: 1867–1879." Ph.D. Diss. U of Colorado, Boulder, 1953.

> Bartlett writes that "Powell's greatest contribution was one of popularization.... [He] created interest in a wild and rugged area" (pp. 352–386).

982. Bernstein, William Samuel. "Lewis Henry Morgan, John Wesley Powell and Henry George: A Study in the Relation Between Nineteenth Century Intellectual Thought and Social Reform." M.A. Thesis. Brown U, 1979.

> Not seen.

983. Darnell, Regna. *The Development of American Anthropology 1879–1920: From the Bureau of American Ethnology to Franz Boas*. Ph.D. Diss. U of Pennsylvania, 1969. Ann Arbor: University Microfilms, 1970.

> Examines the transition period between anthropology as it developed into a profession under Powell's leadership in the BAE and the establishment of Boasian anthropology.

984. Jones, Daniel H. *The Panorama From Point Sublime: John Wesley Powell's "Religion of Science" and the Intellectual Origins of His Arid Lands Reforms*. Ph.D. Diss. Indiana U, 1997. Ann Arbor, MI: UMI Dissertation Services, 1999.

> An examination of Powell's work within the context of America's intellectual transition from "Whig theology" to secularism at the end of the 19th century.

985. Lacey, Michael J. *The Mysteries of Earth-Making Dissolve; A Study of Washington's Intellectual Community and the Origins of American Environmentalism in the Late Nineteenth Century*. Ph.D. Diss. George Washington U, 1979. Ann Arbor, MI: University Microfilms International, 1988.

Not seen.

986. Lapham, Curt. *Major John Wesley Powell: Leadership on the Colorado River, 1869*. M.M.A.S Thesis. U.S. Army Command and General Staff College, 1996. Ft. Belvoir, VA: Defense Technical Information, 1996.

Examines Powell's leadership skills during the 1869 expedition; speculates on what Powell learned during the Civil War.

987. Morris, Lindsey G.. "John Wesley Powell: Scientist and Educator." M.S. Thesis. Illinois State U, 1947.

For annotation, see **#660**.

988. Noelke, Virginia Hull McKimmon. "The Origin and Early History of the Bureau of American Ethnology, 1879–1910." Ph.D. Diss. U of Texas, 1974.

Not seen.

989. Padget, Martin. "Cultural Geographies: Travel Writing in the Southwest, 1869–1897." Ph.D. Diss. U of California, San Diego, 1993.

Examines narratives of exploration and travel with a focus on Powell, Charles F. Lummis, and Helen Hunt Jackson.

990. Steinbacher-Kemp, William. "The Illinois Natural History Society, 1858–1871." M.S. Thesis. Illinois State U, 2000.

Includes Powell's activities in the Society and its demise.

991. Wiseman, Michael. "Don't Pray for Rain—Buy It: The Rational Ordering of Water for Irrigation Through Social Engineering." M.A. Thesis. U of Wyoming, 1988.

Not seen.

992. Zernel, John Joseph. *John Wesley Powell: Science and Reform in a Positive Context*. Ph.D. Diss. Oregon State U, 1983. Ann Arbor, MI: University Microfilms International, 1984.

Study of the intellectual aspects of Powell's survey work.

Chapter 7

Miscellany

A. COLLECTIONS OF PHOTOGRAPHS

993. Denver Public Library, Western History/Genealogy Dept.

Holdings include an extensive collection of photographs documenting the history of the West, including images of Powell and photographs by Jack Hillers. Many images from this collection and the Colorado Historical Society are available online (see **#1020**).

994. National Archives II, Still Pictures Branch. College Park, MD.

(a) National Park Service, RG 79. Still Pictures 1859–1990.

Portraits of Powell and his family, 1859–1868.

(b) Office of the Secretary of the Interior, RG 48. Miscellaneous Photographic Prints.

Photographic portraits of Powell and Interior Dept. officials.

(c) Smithsonian Institution, RG 106. Photographic Negatives of American Delegations and Archaeology of the Southwestern United States.

Over 160 glass negatives and photographs of American Indians, 1870–1907, by John Hillers and other BAE anthropologists. The Smithsonian Institution has retained most of the photographic work of the BAE.

(d) U.S. Geological Survey, RG 57. Records of the Powell Survey, 1869–1879, Photographic Prints and Glass Negatives.

This portion of a massive archive of photographic work of the Old West includes images made 1871–1878 by E. O. Beaman, James Fennemore, and John K. Hillers in Colorado River area. Also includes the Major Joseph Wheat collection of Powell Survey photographic prints, which contain halves of stereographic prints by Beaman, Hillers, and Fennemore. Powell lists each picture by name of tribe and tribe's location in a handwritten catalog.

995. Smithsonian Institution, National Anthropological Archives. Suitland, MD.

Some digital images are available online (see **#1021**).

(a) Portraits of Anthropologists, photographic lot 33, and Portraits of Smithsonian Anthropologists, photographic lot 39.

Portraits of Powell, family, and his house on M Street in Washington.

(b) Glass Negatives of Indians Collected by the Bureau of American Ethnology.

Numbered and cataloged originals and some copy negatives; includes many by John K. Hillers.

(c) Stereographs by the Topological and Geological Survey of the Valley of the Colorado, photographic lots 76–131.

Most from the series Canyons of the Green River and Canyons of the Colorado River of the West, from the 1870s.

996. Smithsonian Institution Archives, Photographic Collection, RU 95. Washington.

Photographs of Smithsonian officials, staff, and scientists.

997. Utah State Historical Society, William Culp Darrah Photograph Collection, ca. 1946–1972. Salt Lake City.

Photographs of Powell, his family, and expeditions relating to *Powell of the Colorado* (**#609**).

B. COLLECTIONS OF MAPS

998. National Archives II. College Park, MD.

(a) Smithsonian Institution, RG 106. Cartographic Records of the Bureau of American Ethnology, 1874–1952.

Contains two published maps prepared by Powell: Indian land cessions for the 1st annual report of the BAE, and Indian linguistic stock map for the 7th annual report.

(b) U.S. Geological Survey, RG57. U.S. Geographical and Geological Survey of the Rocky Mountain Region, 1875–1879.

The Geological Survey of the Territories (Hayden Survey) and the Geological Survey of the Rocky Mountain Region (Powell Survey) operated under the supervision of the Dept. of the Interior and their records are now filed with those of the USGS. Maps representing the work of the USGS during Powell's tenure as director are found in this record group. Cartographic records include published maps of Utah Territory, a few manuscript maps of Utah, a portfolio of published maps of the Black Hills area, and a portfolio of published maps and related records of the geological survey of the Uinta Mountains in Utah.

999. Smithsonian Institution, National Anthropological Archives. Suitland, MD.

(a) Map Collection.

Includes working materials that relate to versions of Powell's *Linguistic Stocks of American Indians North of Mexico* (see **#417**) and other original or annotated maps that concern the work of the BAE.

(b) U.S. National Museum/National Museum of Natural History Records. Dept. of Anthropology, Division of Archeology. Map Collection.

Several maps and manuscripts reflect activities of the BAE.

1000. Utah State Historical Archives, Francis Marion Bishop's 1871 River Maps. Salt Lake City.

Six map sheets covering parts of the Green and Colorado Rivers. Since Bishop sent his official maps to Washington (where they were lost), these are probably his personal copies. See **#770**.

C. AUDIOVISUAL

1001. *99 Days to Survival.* 16 mm. Smithsonian Institution and CBS News, 1973.

E. G. Marshall reads from *Exploration* as modern-day rafters follow Powell's 1869 route. Not seen.

1002. *Grand Canyon: The Courageous First Explorers.* Videocassette. Dir. Christopher Rowley. Prod. National Geographic Society. Dist. Warner Home Video, 1999.

Originally broadcast as an episode in the Great Stories of Adventure series on National Geographic TV.

1003. *John Wesley Powell and the Colorado River.* Filmstrip with audiocassette. Educational Images, 1975.

Not seen.

1004. *John Wesley Powell, Canyon Geologist.* 16mm and videocassette. Dir. Edwin D. McKee. W. A. Palmer Films, 1968.

Also released as videocassette in 1981 by National Audiovisual Center in Washington. Not seen.

1005. *John Wesley Powell Discovery.* Kit. Great American Film Factory, 1975.

Not seen.

1006. *John Powell—Into the Grand Canyon; Chasm Cut by a River.* One-track audiotape. WYNE Radio and New York Board of Education, 1961.

Not seen.

1007. *Lost in the Grand Canyon.* Videocassette. PBS Home Video, 1999.

Originally broadcast as an episode in the PBS series The American Experience (see **#1018**).

1008. *The Real West—John Wesley Powell.* Videocassette. Greystone Communication, 1994.

Not seen.

1009. *River of Stone: The Powell Expedition.* Videocassette. Dir. John Howe. BWE Video, 1997.

Documentary on Powell's 1869 expedition and subsequent use of the Colorado River system. Narrated by Robert Redford. Not seen.

1010. *Ten Who Dared.* Dir. William Beaudine. Disney, 1960.

Feature film on the 1869 expedition. Not seen.

D. INTERNET RESOURCES AND DIGITAL ARCHIVES

The average life span of a website is 100 days. Nevertheless, the following authoritative, stable sites and databases have shown staying power and are thus included.

1011. *American Memory*. Library of Congress. <memory.loc.gov/ammem/amhome.html>.

A treasure-trove of millions of digital images of documents and maps from historical collections relating to the history and culture of the United States. Powell-related items may be found in the following collections:

(a) *Alexander Graham Bell Family Papers at the Library of Congress 1862–1939.*

A selection of correspondence, journals, articles, and more. Includes letters about and to Powell. See **#78**.

(b) *The Evolution of the Conservation Movement 1850–1920.*

From several divisions of the LC, this collection consists of books, manuscripts, maps, prints, and government publications, including *Report on the Lands* and *U.S. Statutes* v. 25, resolution no. 7, "to investigate the practicability of constructing reservoirs for the storage of water in the arid region."

(c) *History of the American West 1869–1920: Photographs from the Collection of the Denver Public Library.*

Selections from the library's Western History and Genealogy Dept. See **#1020**.

(d) *Making of America.*

Primary sources in 19th-century American social history. See **#1019**.

(e) *Map Collections, 1500–2003.*

Digital images from the Geography and Map Division of the LC. Includes maps created for or by Powell or the USGS and BAE during his administration.

(f) *The Nineteenth Century in Print: The Making of America in Books and Periodicals.*

A collection of 19th-century American imprints, mainly 1850–1880, digitized as part of *Making of America* (**#1019**).

1012. *Archival Research Catalog*. U.S. National Archives and Records Administration. <arcweb.archives.gov/research_room/arc/index.html>.

ARC is the online catalog of the holdings of the NARA. The catalog can be searched to view digital images linked to descriptions. Available images include photographs by Hillers and Beaman, and BAE and USGS photographs of Powell and his employees.

1013. *Colorado Plateau Digital Archives*. Northern Arizona U, Cline Library Special Collections and Archives Dept. <www.nau.edu/library/speccoll/index.html>.

This searchable database of images and texts in the Colorado Plateau collection at Cline Library provides access to digital images of photographs by Hillers and documents from the Martin J. Anderson collection (see **#24**).

1014. *J. Willard Marriott Library Digitized Collections*. U of Utah, J. Willard Marriott Library. <www.lib.utah.edu/digital/digcol.html>.

Among items in this collection are Powell's *Exploration* and "The Cañons of the Colorado" in *Scientific American* 1878 (**#262**).

1015. "John Wesley Powell, 1834–1902." *Government Information Connection*. U of North Texas Libraries' Government Documents Dept. <www.library.unt.edu/gpo/powell/powell.htm>.

Biography, images, list of publications, and links to digital images and texts, with access to complete works by Powell from the LC's *American Memory* collection (see **#1011**).

1016. *John Wesley Powell Memorial Museum*. John Wesley Powell Memorial Museum, Page, AZ. <www.powellmuseum.org>.

This site has a biography of Powell and a list of suggested readings.

1017. "John Wesley Powell Photograph Index." *Grand Canyon Photos*. Grand Canyon National Park. <www.nps.gov/grca/photos/powell/pages>.

This site includes digitized photographs of Powell and images from the Colorado River expeditions. Also includes the text from "John Wesley Powell, A Brief Review of His Career," the epilogue from Dellenbaugh's *The Romance of the Colorado River* (**#109**).

1018. "Lost in the Grand Canyon." *American Experience*. PBS/WGBH Educational Foundation. <www.pbs.org/wgbh/amex/canyon/>.

This site supplements the PBS film from The American Experience series (see **#1007**). The site contains timelines, maps, biographical material, and a teacher's guide, with texts from articles by and about Powell in the 1869 *Chicago Tribune* and *Deseret Evening News*.

1019. *Making of America.* U of Michigan. <www.hti.umich.edu/m/moagrp>.

A digital library of primary sources in 19th-century American social history that contains a large and growing collection of books and journal articles from 19th-century imprints, such as the entire *Report on the Geology of the Eastern Portion of the Uintah Mountains* by Powell (**#249**) and Chapter 1 of Beaman's "The Cañon of the Colorado" (**#88**). Part of the LC's *American Memory* collection (see **#1011**).

1020. *The Photography Collection.* Denver Public Library, Western History/ Genealogy Dept. <photoswest.org>

A searchable database of historic photographs from the collections of the Denver Public Library and the Colorado Historical Society that document the history of Colorado and the American West. Includes works of Beaman and Hillers.

1021. *SIRIS Image Gallery.* Smithsonian Institution. <sirismm.si.edu/ sirisimagegallery.html>.

A sampling of archival visual records that are part of the Smithsonian's main catalog (SIRIS). Thousands of images have been digitized and are available by searching the SIRIS Archives, Manuscripts and Photographic Collections catalog at <sirismm.si.edu/siris/ aboutarchives.htm>. Includes images from the Smithsonian Institution Archives and several collections in the NAA: Division of Ethnology Collection; Photographs of American Indians; and Glass Negatives of Indians.

1022. "To the Territories." *The Smithsonian: 150 Years of Adventure, Discovery, and Wonder.* Smithsonian Institution. <www.150si.edu/chap3/ three.htm>.

Chapter 3 of the section titled "Bright Lights, Bold Adventure: 1846–1878" provides an overview of western explorations and their links to the Smithsonian. There is a link to "Beyond the Plateau," which describes Powell's expeditions to the Southwest in the 1860s and 70s.

E. FOR YOUNGER READERS

1023. Anderson, Peter. *Into the Unknown: Major Powell's River Journey.* Highlights from American History. Billings and Helena, MT: Falcon Pr, 1992.

Illustrations by Tom McFarland.

1024. Bruns, Roger. *John Wesley Powell: Explorer of the Grand Canyon.* Historical American Biographies. Springfield, NJ: Enslow Pub, 1997.

Biography for upper-level readers.

1025. Buff, Mary, and Conrad Buff. *The Colorado, River of Mystery.* Los Angeles: Ward Ritchie Pr, 1968.

Human history of the Colorado River with a chapter on Powell.

1026. Calvert, Patricia. *The American Frontier.* Great Lives. New York: Atheneum Bks for Young Readers, 1997.

This collective biography includes Powell.

1027. Carlson, Vada F. *John Wesley Powell: Conquest of the Canyon.* Irvington-on-Hudson, NY: Harvey House, 1974.

A biography, but most of the book is about the 1869 expedition. Drawings by Richard Cuffari.

1028. Copeland, Peter F. *Powell's Colorado River Expedition Coloring Book.* New York: Dover Pub, 1993.

1029. Faber, Doris, and Harold Faber. *Nature and the Environment.* Great Lives. New York: Charles Scribner's, 1991.

Biographies of naturalists and conservationists, including Powell.

1030. Fraser, Mary Ann. *In Search of the Grand Canyon.* Redfeather Books. New York: H. Holt, 1995.

Story of the 1869 expedition. Illustrations by the author.

1031. Frazee, Steve. *First Through the Grand Canyon.* Philadelphia: Winston, 1960.

Written from the first-person point of view.

1032. Gaines, Ann. *John Wesley Powell and the Great Surveys of the American West.* World Explorers. New York: Chelsea House, 1992.

Biography of Powell as geologist.

1033. Hanstein, Otfrid von. *In Den Tälern Des Todes: Die Abenteuerliche Erforschung Der Wunderwelt Am Colorado Durch J. W. Powell.* Jäger Und Forscher. Dresden: Deutsche Buchwerkstätten, 1922. [Rpt. Leipzig: Verlag Deutsche Buchwerkstatten, 1928.]

Aimed at a middle or high school audience, this is a fictionalized account of *Exploration.* In German.

1034. Harmer, Mabel. "Adventure on the Green River: John Wesley Powell's 1869 River Expedition." *Beehive History* 1 (1975): 18–22.

Retells the Green River portion of the 1869 expedition.

1035. Jackson, Paul. "Conquerors of the Canyon." *Coronet* 26 (September 1949): 121–124.

The story of the 1869 Colorado River expedition. Based on a broadcast of *You Are There* (CBS), originally written by Irve Tunick.

1036. Kent, Deborah, and Kathryn A. Quinlan. *Extraordinary People With Disabilities.* New York: Children's Pr, 1996.

Profiles seven dozen people throughout history with various physical or mental disabilities, including Powell.

1037. Maurer, Richard. *The Wild Colorado: The True Adventures of Fred Dellenbaugh, Age 17, on the Second Powell Expedition Into the Grand Canyon.* New York: Crown Pub, 1999. [Rpt. New York: Scholastic Inc, 2000.]

1038. Maynard, Charles William. *John Wesley Powell, Soldier, Scientist, and Explorer.* Famous Explorers of the American West. New York: PowerKids Pr, 2003.

For very young readers.

1039. Place, Marian Templeton. *John Wesley Powell; Canyon's Conqueror.* Boston: Houghton Mifflin, 1963.

Illustrated by Harve Stein. See also **#1048**.

1040. Powell, John Wesley. *The Diary of John Wesley Powell: Conquering the Grand Canyon.* Ed. Connie and Peter Roop. In My Own Words. New York: Benchmark Bks, 2001.

Extracts from *Exploration*, abridged and illustrated for middle readers.

1041. Powell, John Wesley. "In Search of a River's End." *Cobblestone* 1.6 (1980): 11–19.

Excerpts from *Exploration* in a magazine for older readers.

1042. Prescott, Jerome. *100 Explorers Who Shaped World History*. San Mateo, CA: Bluewood Bks, 1996.

Includes Powell and Hayden.

1043. Ross, Michael Elsohn. *Exploring the Earth With John Wesley Powell.* Naturalist's Apprentice. Minneapolis, MN: Carolrhoda Bks, 2000.

Biography with activities for children related to geology and earth sciences. Illustrations by Wendy Smith.

1044. Smith, Vera Lee. "Powell Journey Bares Canyon Secrets." *National Geographic School Bulletin*. 47.30 (1969 May 5): 474–475.

From an issue devoted to "Exploring the Unknown," juxtaposing Powell's journey to Grand Canyon with astronauts to the moon.

1045. Summers, Richard Aldrich. *Conquerors of the River*. New York: Oxford U Pr, 1939.

Highly fictionalized account of first Grand Canyon expedition. Illustrated by Donald McKay.

1046. Ullman, James Ramsey. *Down the Colorado With Major Powell.* North Star Books, 17. Boston: Houghton Mifflin, 1960.

Illustrated by Nicholas Eggenhofer.

1047. Washburn, Patrick S. "John Wesley Powell: Explorer of the Grand Canyon." *Boys' Life*. 84.3 (1994): 28.

1048. White, Dale [Marian Templeton Place]. *John Wesley Powell: Geologist-Explorer*. New York: Julian Messner, 1958.

Dale White is a pseudonym for Marian T. Place (see **#1039**). This is a highly fictionalized biography.

1049. Wibberley, Leonard. *Wes Powell, Conqueror of the Grand Canyon*. New York: Ariel Bks, 1958.

Biography for older readers.

F. JOHN WESLEY POWELL IN FICTION, POETRY, AND OTHER TEXTS

1050. Belknap, Buzz, and John Wesley Powell. *Grand Canyon River Guide.* New York: Canyonlands Pr, 1969.

At head of title: Powell Centennial. Includes condensed quotations from *Exploration.*

1051. Gill, Ardian. "*The River Is Mine": John Wesley Powell's 1869 Exploration of the Green and Colorado Rivers and the Grand Canyon.* New York: Local Color Pr, 2002.

This novel recreates the 1869 expedition, primarily from the point of view of George Bradley. Gill relies on published diaries and letters of members of the exploring party.

1052. Hood, Charles. *Red Sky, Red Water: Powell on the Colorado.* Tucson: SUN/Gemini Pr, 1990.

A poem.

1053. *The Spirit of the Earth: Selections From the John Wesley Powell Collection of Pueblo Pottery at Illinois Wesleyan University, 22 October 1993 to 9 January 1994.* Urbana: Krannert Art Museum and Kinkead Pavilion, U of Illinois at Urbana-Champaign, 1993.

Exhibit catalog for the IWU John Wesley Powell Collection of Traditional Pueblo pottery.

1054. Vernon, John. *The Last Canyon.* Boston: Houghton Mifflin, 2001.

A novel of the 1869 Colorado River expedition.

Author Index

Numbers refer to bibliographic entries.

Q

R

S

Subject Index

Numbers refer to bibliographic entries.

J

K

L

M

N

T

U

Title Index of Works by John Wesley Powell

Numbers refer to bibliographic entries.

B

C

D

E

F

G

H

I

J

L

M

N

R

S

ABOUT THE AUTHOR

Marcia L. Thomas is Associate Professor at Illinois Wesleyan University, where she is also Director of Technical Services for the Ames Library.

Recent Titles in
Bibliographies and Indexes in American History

Native North American Shamanism: An Annotated Bibliography
Shelley Anne Osterreich, compiler

Guide to Native American Ledger Drawings and Pictographs in United States Museums, Libraries, and Archives
John R. Lovett, Jr., and Donald L. DeWitt, compilers

Native America and the Evolution of Democracy: An Annotated Bibliography
Bruce E. Johansen, compiler

The Kickapoo Indians, Their History and Culture: An Annotated Bibliography
Phillip M. White, compiler

The American Settlement Movement: A Bibliography
Domenica M. Barbuto, compiler

Reconstruction in the United States: An Annotated Bibliography
David A. Lincove, compiler and annotator

Books on Early American History and Culture, 1991–1995: An Annotated Bibliography
Raymond D. Irwin

Peyotism and the Native American Church: An Annotated Bibliography
Phillip M. White

A Comprehensive Catalogue of the Correspondence and Papers of James Monroe, Volume I
Daniel Preston

A Comprehensive Catalogue of the Correspondence and Papers of James Monroe, Volume II
Daniel Preston

Books on Early American History and Culture, 1986–1990: An Annotated Bibliography
Raymond D. Irwin

American Indian and African American People, Communities, and Interactions: An Annotated Bibliography
Lisa Bier